Springer INdAM Series

Volume 9

Editor-in-Chief

V. Ancona

Series Editors

P. Cannarsa
C. Canuto
G. Coletti
P. Marcellini
G. Patrizio
T. Ruggeri
E. Strickland
A. Verra

For further volumes:
http://www.springer.com/series/10283

Françoise Dal'Bo • Marc Peigné •
Andrea Sambusetti

Editors

Analytic and Probabilistic Approaches to Dynamics in Negative Curvature

 Springer

Editors
Françoise Dal'Bo
IRMAR
Université de Rennes 1
Rennes
France

Marc Peigné
Lab. de Mathématiques et Physique
Université François Rabelais
Tours
France

Andrea Sambusetti
Dipartimento di Matematica
Sapienza - Università di Roma
Roma
Italy

ISSN 2281-518X   ISSN 2281-5198 (electronic)
Springer INdAM Series
ISBN 978-3-319-04806-2   ISBN 978-3-319-04807-9 (eBook)
DOI 10.1007/978-3-319-04807-9
Springer Cham Heidelberg New York Dordrecht London

Library of Congress Control Number: 2014943952

Preface

The ergodicity of the geodesic flow $(g_t)_{t\in\mathbb{R}}$ with respect to the Liouville measure m on the unit tangent bundle T^1S of a compact surface (or of finite area) with curvature -1 was originally proved by G. A. Hedlund and E. Hopf in the 1930s, by a simple and enlightening idea, the *Hopf argument*: any measurable set invariant under the geodesic flow is actually invariant by the stable and unstable distributions of the flow. A different and elegant argument, holding in the more general setting of symmetric spaces, and based on some commutation relations between diagonal and unipotent matrices, was used in the 1950s by F. I. Mautner. However, the dynamical approach initiated by Hopf was so robust to hold not only for the geodesic flow of compact surfaces with negative, constant curvature, but also for the geodesic flow of general n-manifolds with variable negative curvature, and even for a much larger class of dynamical systems, now called *hyperbolic*.[1] The door to more complex situations and questions was then open!

By the ergodicity of $(g_t)_{t\in\mathbb{R}}$, we knew that for any observable $\varphi \in \mathbb{L}^1(m)$, the quantity $\dfrac{1}{t} \displaystyle\int_0^t \varphi(g_s \cdot v)ds$ converges m-almost surely towards the integral of φ with respect to m; the question of the optimal normalization of these sums, for observables φ with mean 0, naturally arised. The investigation of the convergence of properly normalized ergodic sums towards a Gaussian law needed the development of new, important tools; by a control of the speed of equirepartition of suitable means towards the Liouville measure, in 1960, Y. G. Sinaï proved a Central Limit Theorem for the geodesic flow of compact manifolds M with constant negative curvature, for a wide class of regular observables φ:

---

[1] It is worth to stress that the horocycle flow is far to be *hyperbolic*; nevertheless, the intertwining relations with the geodesic flow lead quite naturally to its unique ergodicity. In this regard, let us cite at least the original works of H. Furstenberg and B. Marcus, and a recent work of Y. Coudène proving, along the lines of Marcus' dynamical proof, the unique ergodicity of the horocycle flow associated to an Anosov flow with one dimensional orientable strong stable distribution. Here again we see an argument *à la Hopf*, based on the local product structure of the unit tangent bundle induced by the stable and unstable foliations of the geodesic flow.

There exists a constant $\sigma := \sigma(\phi) > 0$ such that, for any $A \in \mathbb{R}$

$$\lim_{t \to +\infty} m \left\{ x : \frac{1}{t} \int_0^t \varphi(g_s \cdot v)ds - \bar{\phi} \leq A \cdot \frac{\sigma}{\sqrt{t}} \right\} = \frac{1}{\sqrt{2\pi}} \int_{-\infty}^A e^{-s^2/2}ds$$

where $\bar{\phi} := \int_{T^1 M} \phi(v)m(dv)$.

During the 1960s and 1970s, another approach, using codings given by Markov partitions and tools coming from thermodynamic formalism, allowed several authors to describe the stochastic behavior of a larger class of dynamical systems, the so-called *Anosov systems*. For instance, a Central Limit Theorem was obtained by M. Ratner for special flows built over subshifts of finite type and satisfying an uniform exponential mixing; the method relied on the local expansion of the dominant eigenvalue of a family of operators, the so-called *transfer* or *Ruelle* operators with potential corresponding to the roof function of the special flow. The remarkable point of this approach is again its flexibility, being valid for the geodesic flow of general compact, negatively curved manifolds.

Let us mention here another original approach to the Central Limit Theorem for the geodesic flow, due to Y. Le Jan and J. Franchi, based on the comparison between geodesics and trajectories of the hyperbolic Brownian motion. The basic idea is to notice that the integral of some regular observable φ along a geodesic coincides with the integral of a closed 1-form along the stable leaf defined by this geodesic and the corresponding horocycles at $+\infty$; then, changing the integration path, the geodesic may be replaced by a Brownian motion path on this leaf. This approach can be used to extend the Ratner-Sinai theorem to noncompact manifolds of constant curvature, and does not require any coding (which is difficult to obtain in the finite volume case); on the other hand, it needs some refined results in potential theory, forcing in particular the curvature to be constant, which was not the case in Ratner's approach. Indeed, as for the coding method, an important role in the proof is played by a spectral gap argument, yielding a potential operator for the geodesic flow.

Still in the 1960s, the classical counting problem for closed geodesics on a negative curved manifold was also deeply investigated, through several other approaches, adapted to each context. For instance, for compact surfaces of constant curvature -1, H. Huber obtained in 1959 the main term of the asymptotic behaviour of the number of closed geodesic with length smaller than t, for $t \to +\infty$. G. Margulis, S. J. Patterson, D. Sullivan and others proposed quite different techniques to approach this counting problem in negative, variable curvature, for compact and non-compact manifolds. The great challenge was: first, to construct the unique measure of maximal entropy (the so-called *Bowen-Margulis measure*) for the flow $(g_t)_{t \in \mathbb{R}}$ restricted to its non-wandering set; second, to describe the expansion properties of the associated conditional measures with respect to the stable and unstable foliations. A complete answer, in a very general setting, was given in 2003 by T. Roblin in his seminal work in *Bulletin de la SMF*, for any discrete group of isometries Γ of a CAT(-1) space with finite Bowen-Margulis measure, provided that Γ has a non-arithmetic length spectrum:

There exists a constant $C(x, y) > 0$ such that

$$N_\Gamma(x, y|R) := \sharp\{\gamma \in \Gamma \; : \; d(x, \gamma \cdot y) \leq R\} = (C(x, y) + o(R)) \, e^{\delta_\Gamma R}$$

where δ_Γ is the Poincaré critical exponent of the group.

On the other hand, the counting problem was also investigated for Anosov flows, for instance by D. Ruelle, R. Bowen, M. Pollicott et al., via transfer operators $(\mathscr{L}_s)_{s \in \mathbb{R}}$ and the corresponding dynamical Zeta functions. The fact that the dominating eigenvalue of $\mathscr{L}_s$, restricted to some suitable functional space, is an isolated point of the spectrum (the one of maximal modulus) is a key argument of all these works,[2] and is closely related to the mixing property of the Anosov flow. This is the approach which led Ratner to the Central Limit Theorem; using techniques from renewal theory for Markov random walks, this method also yields precise estimates of the main asymptotic term of the counting functions considered above, thus giving another illustration of the stochastic behavior of the geodesic flow in negative curvature.

While the situation concerning the asymptotic dominant term of the counting function of closed geodesics (and closed paths) in negative curvature is now well known, the question of their asymptotic expansions is always quite open. During the 1970s, for hyperbolic surfaces $S = \Gamma \backslash \mathbb{H}^2$ of finite area, Patterson obtained such expansions, with control of the error term, by a method based on the Selberg trace formula related to the hyperbolic Laplacian:

Let $0 = \lambda_0 > \lambda_1 \geq \cdots \geq \lambda_q > -\frac{1}{4}$ be the eigenvalues greater than $-\frac{1}{4}$ of the hyperbolic Laplacian, with corresponding eigenfunctions $\phi_1, \cdots, \phi_q$; then, for all $x, y \in \mathbb{H}^2$

$$N_\Gamma(x, y|R) = \frac{\pi}{area(S)} e^R + \sqrt{\pi} \sum_{k=1}^{q} \frac{(s_k - 3/2)!}{s_k!} \phi_k(x)\phi_k(y)e^{s_k R} + O(e^{\frac{3}{4}R})$$

where $s_k := \frac{1}{2} + \sqrt{\frac{1}{4} + \lambda_k}$.

Further extensions of this result have followed, first of all for finite volume hyperbolic manifolds; but the question is much more complex in variable curvature, or when the volume is infinite, even for geometrically finite manifolds. This question is closely related to the control of the speed of mixing for the flow; D. Dolgopyat's work on decay of correlation in Anosov flows, at the end of the 1990s, led several authors to an upper bound for the error term in the asymptotic expansion. For instance, for any compact, negatively curved surface S with fundamental group Γ, M. Pollicott and R. Sharp showed:

---

[2]Whatever method we adopt, a spectral gap property appears somewhere! In the transfer operator approach, the general paradigm is: first, proving the existence of a unique $s_0 \in \mathbb{R}$ such that the spectral radius of $\mathscr{L}_{s_0}$ is equal to 1; then, deducing the formulas for the local expansion at s_0 of the dominant eigenvalue of the $\mathscr{L}_s$.

For all x, y in the universal covering $\hat{S}$, there exists a constant $C(x, y) > 0$ such that

$$N_\Gamma(x, y|R) = C(x, y)e^{\delta_\Gamma R} + O(e^{\lambda R}).$$

for some $0 < \lambda < \delta_\Gamma$ which does not depend on x, y.

However, if we look for an analogous of Patterson's result, we need a more flexible operator, whose infinitesimal generator replaces the hyperbolic Laplacian and whose spectrum can be controlled. This domain is presently very active and many progresses have been obtained in the last years.

The two texts presented in this book are independent and concern directly these two topics: the Central Limit Theorem (CLT) and the counting problem.

Martingale methods for hyperbolic dynamical systems, by Stéphane Le Borgne, focuses on the CLT for a large class of hyperbolic systems, via martingales theory. Based on Gordin's decomposition, this method requires some information about the speed of equirepartition of some means on the corresponding unstable foliation. In the case of the geodesic flow, it corresponds to a quantitative control of the ergodicity of the unstable horocycle flow.[3] The approach by martingales presented by Leborgne can be used in several situations, even for weakly hyperbolic flows: e.g., diagonal flows on compact quotients of $SL(d; \mathbb{R})$.

The second text of the book, *A Semiclassical Approach for the Ruelle-Pollicott Spectrum of Hyperbolic Dynamics*, by Frédéric Faure and Masato Tsujii, is related to the counting problem via thermodynamic formalism. In this situation, the transfer operator associated to the dynamic replaces the hyperbolic Laplacian for the geodesic flow in constant curvature -1, and the question is to describe the structure of its spectrum. For instance, when the dynamical system is a contact Anosov flow, the *Ruelle-Pollicott spectrum* of its generator has a structure in vertical bands, and the trace formula of Atiyah-Bott then leads to an asymptotic expansion of the counting function of periodic orbits; one may also define a Zeta function which generalizes the Selberg Zeta function in the case of constant curvature and which has similar properties: i.e. its zeroes lie, asymptotically, on a vertical line. In this text, the purpose of the authors is not to analyze the asymptotic development of N_Γ; however, they show how to recover the dominant term, and their accurate description of the spectrum will certainly lead, in further study, to more refined counting formulas for the periodic orbits of the flow.

Rennes, France   Françoise Dal'Bo
Tours, France   Marc Peigné
Roma, Italy   Andrea Sambusetti
December 2013

---

[3]We notice that also Sinai's original approach to the CLT needed such a quantitative information, as mentioned before.

Acknowledgements

This book stems from a series of lectures given at the Indam Workshop "Geometric, Analytic and Probabilistic Approaches to Dynamics in Negative Curvature", organized by the Editors in Rome, 13–17 May 2013.

The workshop was mainly supported by INdAM, with contributions by the GdR Platon "Géométrie, Arithmétique et Probabilités", the Sapienza Università di Roma, the PRIN project "Spazi di Moduli e Teoria di Lie", the FIRB programme "Geometry and Topology of Low-dimensional Manifolds" and the ERC project "Macroscopic Laws and Dynamical Systems". The Editors express their gratitude to all the contributors, especially to INdAM for the precious collaboration of its staff.

We would like to thank the four lecturers of the meeting: Frédéric Faure, Jacques Franchi, Stéphane Le Borgne and Amie Wilkinson. Moreover, we are indebted to the Scientific Committee, which was composed by François Ledrappier, Carlangelo Liverani and Gabriele Mondello, for their encouragement and invaluable advice.

Contents

1 Martingales in Hyperbolic Geometry.. 1
Stéphane Le Borgne

2 Semiclassical Approach for the Ruelle-Pollicott Spectrum
of Hyperbolic Dynamics... 65
Frédéric Faure and Masato Tsujii

Index.. 137

Chapter 1
Martingales in Hyperbolic Geometry

Stéphane Le Borgne

Abstract The famous De Moivre-Laplace theorem states the convergence toward a gaussian law of $\sum_{j=0}^{n-1} Y_j / \sqrt{n}$ when the Y_i are independent, centered, identically distributed random variables in L^2. This result is usually named Central Limit Theorem (*CLT*). The convergence still holds in some non independent cases (Markov chains, α- or ϕ-mixing processes, martingales,...). Here we are interested in stationary processes defined by regular functions on regular hyperbolic systems. We show how the martingales formalism is well fitted to get the CLT in such situations. First we prove a few results on martingales and present Gordin's method. Then we employ the method for two toy model dynamical systems: the angle doubling on the circle and the cat map. After what, the presented ideas are applied to more general dynamical systems (among which 1960 Sinaï's example and some other geodesic flows on hyperbolic manifolds). We stress the importance of the equirepartition of some submanifolds and explain how this can be related to the mixing properties of the system. As an example of application, we study certain asymptotic properties of random walks on $\mathbb{R}^d$ driven by a hyperbolic system.

1.1 Introduction

As its name tells us the central limit theorem is a very important result of the probability theory. It has many practical and theoretical consequences. First established for the sequences of independent identically distributed random variables (De Moivre-Laplace theorem), it has been then proved in other situations (Markov chains, α- or ϕ-mixing stationary processes, martingales,...) where the independence condition is weakened.

S. Le Borgne (✉)
UFR de Mathématiques, Université de Rennes 1, Campus de Beaulieu, 35042 Rennes, France
e-mail: stephane.leborgne@univ-rennes1.fr

F. Dal'Bo et al. (eds.), *Analytic and Probabilistic Approaches to Dynamics in Negative Curvature*, Springer INdAM Series 9, DOI 10.1007/978-3-319-04807-9_1,
© Springer International Publishing Switzerland 2014

The problem of the convergence towards a Gaussian law of the suitably normalized ergodic sums for a dynamical system has been studied in the 1940s [19]. In the 1960s in Russia important tools have been created that allow to address the problem for a class of dynamical systems usually called hyperbolic.

Since the paper of Sinaï [46] published in 1960, the proof of the central limit theorem (CLT) in the hyperbolic dynamical systems has been the subject of many works. For the Anosov systems the first proofs were often based on codings given by Markov partitions (see for example [2, 24, 42, 47]). More recently other techniques have been devised (see for example [5, 16, 26, 38, 44]). Different methods have also been introduced to deal with more general cases: non compactness, weak hyperbolicity ([18, 22, 36, 49],...).

A stationary real process is defined by a dynamical system, that is a triplet (X, T, μ) where $T : X \to X$ is a measurable transformation of a measurable space X preserving the probability measure μ: when φ is a measurable real function on X, the functions $Y_k = \varphi \circ T^k$ form a stationary sequence of random variables on the probability space (X, μ). If (X, T, μ) is ergodic and φ is integrable, then the averages $\sum_{j=0}^{n-1} Y_j / n$, μ-almost surely tends to $\int_X \varphi \, d\mu$ (this is the Birkhoff theorem). One says that a centered function φ *satisfies the CLT* if there exists a *strictly positive* number σ such that

$$\mu\{x \,;\, \frac{1}{\sqrt{n}} \sum_{j=0}^{n-1} \varphi(T^j x) \in A\} \longrightarrow \frac{1}{\sqrt{2\pi}\sigma} \int_A \exp(\frac{-t^2}{2\sigma^2}) dt.$$

Here, we will be interested in stationary processes defined by regular functions on regular hyperbolic systems. We will show how the martingales formalism is well fitted to get the CLT in such situations. We insist on the geodesic flow on hyperbolic manifolds (other examples are briefly mentioned).

The martingales method, developed by Billingsley, Ibragimov then Gordin [21] give more precise results than the CLT, such as the invariance principle and can be applied in cases where no Markov partition exists (see for example the papers [17, 33, 34, 37, 39, 49]).

We will show how this method works in algebraic different cases. We will insist on the few simple ideas that can be used. We will try to give complete proofs... that will often be based on celebrated strong results which would necessitate many pages to be detailed. In the same spirit we won't give the more general statement: for example, we will most of the time restrict ourselves to bounded functions and avoid every technicality induced by some lack of regularity of the functions involved.

First we will prove a few results on martingales and present Gordin's method. Then we will employ the method for two toy model dynamical systems: the angle doubling on the circle and the cat map. After what, the presented ideas will be applied to more general dynamical systems (among which 1960 Sinaï's example and some other geodesic flows on hyperbolic manifolds). We will stress the importance of the equidistribution of some submanifolds and explain how this can be related to the mixing properties of the system. As an example of application, we will study certain asymptotic properties of random walks on $\mathbb{R}^d$ driven by a hyperbolic system.

Most of the material presented here is very classical. The techniques are applied to several examples beginning by the simplest. Proofs are not always repeated. The following points might be more original. In Sect. 1.4.1 (Example 5) we obtain a CLT for the frame bundle flow: this is not an Anosov flow and the theorem of Ratner doesn't apply in this case. In Sect. 1.5.2 (Example 8) we show how one can modify the martingale technique to get CLT for the geodesic flow along a non arithmetic sequence of times (in particular we are not in the stationary case anymore). In Sect. 1.4.4, we show how the CLT can provide information on the behaviour of the geodesic flow defined on infinite volume surfaces fibered over a finite volume one with $\mathbb{Z}^d$ fibers. The case of a non compact finite volume base may be less classical than the one of a compact base (it is known that completely different behaviours may occur [43]).

All the examples treated here are algebraic: the actions we deal with are defined on homogeneous spaces equipped by the associated Haar measure (the constant curvature case in hyperbolic geometry). Other cases can also be treated by the techniques presented here. For example, for a contact Anosov flow, Liverani has proved in [38] that the system is exponentially mixing for the Liouville measure. Using the result of Anosov [1] asserting the absolute continuity of the holonomy map we can reason like in Sect. 1.4 and get the CLT.[1]

The values of the constant C that appears often in the text may change from line to line.

1.2 Martingales and Central Limit Theorem in Dynamical Systems

1.2.1 The De Moivre-Laplace Theorem

Definition 1.1. Let (Y_n) be a sequence of real random variables and Z another one. One says that Y_n tends to Z in distribution if, for every continuous function φ with compact support, one has

$$\mathbb{E}(\varphi(Y_n)) \to_{n\to\infty} \mathbb{E}(\varphi(Z)).$$

Theorem 1.1. *Let (Y_n) be a sequence of real random variables and Z another one. The sequence Y_n tends to Z in distribution if, for every t, one has*

$$\mathbb{E}(\exp(itY_n)) \to_{n\to\infty} \mathbb{E}(\exp(itZ)).$$

---

[1]Remark that the CLT has been established by Ratner for every Anosov flow in [42] even if the flow is not rapidly mixing.

If Z is a centered gaussian variable with variance σ^2 then its characteristic function is $\mathbb{E}(\exp(itZ)) = \exp(-\sigma^2 t^2/2)$. The De Moivre-Laplace theorem asserts that normalized sums of independent square integrable variables converge in distribution to a gaussian variable.

Theorem 1.2. *Let (Y_k) be a sequence of random variables, independent, identically distributed with expectation $\mathbb{E}(Y_1)$ and variance $\sigma^2 > 0$. Then the sequence*

$$\frac{1}{\sqrt{n}}(Y_1 + Y_2 + \ldots + Y_n - n\mathbb{E}(Y_1))$$

converges in distribution toward a centered gaussian law of variance σ^2.

We write S_n for the sum $Y_1 + \ldots + Y_n$.

Proof. To be as simple as possible we will only consider bounded centered variables. One has:

$$\mathbb{E}(\exp(itS_n/\sqrt{n})) = \prod_{k=1}^{n} \mathbb{E}(\exp(itY_k/\sqrt{n})) = \mathbb{E}(\exp(itY_1/\sqrt{n}))^n.$$

We the use Taylor expansion: $\exp(itY_1/\sqrt{n}) = 1 + itY_1/\sqrt{n} - t^2Y_1^2/2n + O(1/n^{3/2})$. This gives

$$\mathbb{E}(\exp(itY_1/\sqrt{n})) = 1 + it\frac{\mathbb{E}(Y_1)}{\sqrt{n}} - t^2\frac{\mathbb{E}(Y_1)^2}{2n} + O(\frac{1}{n^{3/2}}) = 1 - \frac{\sigma^2 t^2}{2n} + O(\frac{1}{n^{3/2}})$$

and

$$\mathbb{E}(\exp(itY_1/\sqrt{n}))^n = (1 - \frac{\sigma^2 t^2}{2n} + O(\frac{1}{n^{3/2}}))^n \to_{n\to\infty} \exp(-\sigma^2 t^2/2). \qquad \square$$

Now we will see that we can easily relax the condition of independence of the theorem. Let us give another way of rewriting the exponential that will be used below.

Lemma 1.1. *For $|x| < 1$, the following equality defines a function φ:*

$$\exp(ix) = (1 + ix)\exp\left(-\frac{x^2}{2} + \varphi(x)\right),$$

and, for every $|x| < 1$, one has $|\varphi(x)| < \frac{|x|^3}{2}$.

Proof. We use the function Log defined on the complex plane by:

$$\text{Log}(1 + z) = \sum_{k=0}^{+\infty}(-1)^k\frac{z^{k+1}}{k + 1},$$

for $|z| < 1$. One has $\exp(\mathrm{Log}(z)) = z$, for $|z| < 1$. The error bound for alternating series by the modulus of the first term of the remainder gives the following:

$$\left| \mathrm{Log}(1 + ix) - ix - \frac{x^2}{2} \right| \le |\frac{x^3}{3} + i\frac{x^4}{4}| \le \frac{|x|^3}{2}.$$

This leads to

$$\mathrm{Log}\left(\frac{\exp(ix)}{1 + ix} \right) = ix - \mathrm{Log}(1 + ix) = -\frac{x^2}{2} + \varphi(x)$$

with $|\varphi(x)| \le \frac{|x|^3}{2}$. Taking the image by exp one obtains the claim.   □

Let (Y_k) be a sequence of variables. We want to study the expectation

$$\mathbb{E}(\exp(itS_n/\sqrt{n})).$$

Using the lemma we can transform this expression

$$\mathbb{E}(\exp(itS_n/\sqrt{n})) = \mathbb{E}\left(\prod_{k=1}^{n} \exp(itY_k/\sqrt{n}) \right)$$

$$= \mathbb{E}\left(\prod_{k=1}^{n}(1 + it\frac{Y_k}{\sqrt{n}}) \prod_{k=1}^{n} \exp\left(-\frac{t^2 Y_k^2}{2n} + \varphi(\frac{tY_k}{\sqrt{n}}) \right) \right)$$

$$= \mathbb{E}\left(\prod_{k=1}^{n}(1 + it\frac{Y_k}{\sqrt{n}}) \exp\left(\sum_{k=1}^{n}\left(-\frac{t^2 Y_k^2}{2n} + \varphi(\frac{tY_k}{\sqrt{n}}) \right) \right) \right).$$

We now place ourselves in the case where Y_k is a bounded sequence generated by an ergodic dynamical system: there exist $(\Omega, \mathbb{P})$ a probability space, T a measure preserving transformation of Ω, a function f on Ω such that $Y_k = f \circ T^k$. Because of the ergodicity

$$\sum_{k=1}^{n} \frac{Y_k^2}{2n}$$

converges almost surely to $\mathbb{E}(\frac{Y_1^2}{2})$, and as f is bounded

$$\left| \sum_{k=1}^{n} \varphi(\frac{tY_k}{\sqrt{n}}) \right| \le \frac{t^3 \|f\|_\infty^3}{\sqrt{n}}$$

when n is large enough so that $\frac{|t|\|f\|_\infty}{\sqrt{n}} < 1$. So we can split the preceding quantity as follows:

$$\mathbb{E}(\exp(itS_n/\sqrt{n})) = \mathbb{E}\left(\prod_{k=1}^{n}(1 + it\frac{Y_k}{\sqrt{n}})e^{-t^2\frac{\mathbb{E}(Y_1^2)}{2}}\right)$$

$$+\mathbb{E}\left(\prod_{k=1}^{n}(1 + it\frac{Y_k}{\sqrt{n}})\left(e^{\sum_{k=1}^{n}\left(-t^2\frac{Y_k^2}{2n}+\varphi(\frac{tY_k}{\sqrt{n}})\right)} - e^{-t^2\frac{\mathbb{E}(Y_1^2)}{2}}\right)\right).$$

As the variables Y_k are bounded by $\|f\|_\infty$ the modulus of $\prod_{k=1}^{n}(1+it\frac{Y_k}{\sqrt{n}})$ is smaller than $\exp(\frac{t^2\|f\|_\infty^2}{2})$. The theorem of Lebesgue thus implies that

$$\mathbb{E}\left(\prod_{k=1}^{n}(1 + it\frac{Y_k}{\sqrt{n}})\left(e^{\sum_{k=1}^{n}\left(-t^2\frac{Y_k^2}{2n}+\varphi(\frac{tY_k}{\sqrt{n}})\right)} - e^{-t^2\frac{\mathbb{E}(Y_1^2)}{2}}\right)\right)$$

tends to 0.

Definition 1.2. A discrete time process of square integrable random variables (Y_k) is said to be a sequence of *differences of martingale* with respect to an increasing sequence of σ-algebras $(\mathscr{F}_n)$ if it satisfies the following conditions:

(i) Y_n is $\mathscr{F}_n$-measurable.
(ii) $\mathbb{E}(Y_n|\mathscr{F}_{n-1}) = 0$, for every n.

Definition 1.3. A discrete time process of square integrable random variables (Y_k) is said to be a sequence of *differences of a reversed martingale* with respect to an decreasing sequence of σ-algebras $(\mathscr{A}_n)$ if it satisfies the following conditions:

(i) Y_n is $\mathscr{A}_n$-measurable.
(ii) $\mathbb{E}(Y_n|\mathscr{A}_{n+1}) = 0$, for every n.

In these cases the expectations

$$\mathbb{E}(\prod_{j} Y_{k_j})$$

vanish when the k_j are different. For example for the differences of a martingale when the sequence $(k_j)_j$ is strictly increasing:

$$\mathbb{E}(\prod_{j}^{r} Y_{k_j}) = \mathbb{E}(\mathbb{E}(\prod_{j}^{r} Y_{k_j}|\mathscr{A}_{k_r-1})) = \mathbb{E}(\prod_{j}^{r-1} Y_{k_j}\mathbb{E}(Y_{k_r}|\mathscr{A}_{k_r-1})) = 0.$$

It means in particular that for differences of martingale (or reversed martingale) we have

$$\mathbb{E}\left(\prod_{k=1}^{n}(1 + it\frac{Y_k}{\sqrt{n}})\right) = 1.$$

As a consequence if Y_k is a sequence of bounded differences of a stationary ergodic martingale we get

$$\mathbb{E}(\exp(itS_n/\sqrt{n}))$$

$$= e^{-t^2\frac{\mathbb{E}(Y_1^2)}{2}} + \mathbb{E}\left(\prod_{k=1}^{n}(1 + it\frac{Y_k}{\sqrt{n}})\left(e^{\sum_{k=1}^{n}\left(-t^2\frac{Y_k^2}{2n} + \varphi(\frac{tY_k}{\sqrt{n}})\right)} - e^{-t^2\frac{\mathbb{E}(Y_1^2)}{2}}\right)\right)$$

$$\rightarrow_{n\to\infty} e^{-t^2\frac{\mathbb{E}(Y_1^2)}{2}}$$

This proves the CLT for these sequences.

Using the dynamical systems terminology we can state the following theorem:

Theorem 1.3. *Let* $(X, \mathcal{B}, \mu, T)$ *be an invertible dynamical system, f a bounded measurable function and $\mathcal{A}$ a sub-σ-algebra of $\mathcal{B}$ such that:*

(i) $\mathcal{A} \subset T\mathcal{A}$.
(ii) f *is $\mathcal{A}$-measurable.*
(iii) $\mathbb{E}\left(f|T^{-1}\mathcal{A}\right) = 0$.
(iv) $\int f^2 d\mu = \sigma^2(f) > 0$.

Then the function f satisfies the CLT.

Proof. The hypothesis insure that the sequence $(T^k f)$ is a sequence of differences of a reversed martingale with respect to the decreasing filtration $(T^{-k}\mathcal{A})$:

$$\mathbb{E}\left(T^k f|T^{-k-1}\mathcal{A}\right) = \mathbb{E}\left(f|T^{-1}\mathcal{A}\right) \circ T^k = 0. \qquad \square$$

Since the works of Billingsley and Ibragimov many other results on differences of martingales have been obtained (see [9,25] for example). Random variables don't have to be bounded. The result still holds for L^2 variables.

A dynamical system being given, are there functions f and filtrations $\mathcal{A}_n$ such that the sequence $(T^n f)$ is a sequence of differences of martingale? As we will see the answer is yes in hyperbolic dynamical systems. But it must be emphasized that the question which we are interested in is not: Are there functions which satisfies the CLT? The answer is always yes [11,48]. Our question is: Do regular functions satisfy the CLT? And in general the regular functions do not generate a sequence of differences of martingale with respect to any filtration.

We have to relax the hypothesis of martingale. It can be done by making use of a very simple idea first introduced by Gordin [21]: many stochastic properties of ergodic sums associated to a function f are still valid for functions cohomologous to f. One can enlarge the applicability of the preceding result.

1.2.2   Example 1: The Angle Doubling

Let (X, T, μ) defined as follows: $X = [0, 1]$,

$$T : x \mapsto 2x \bmod 1,$$

μ the Lebesgue measure on $[0, 1]$. Let φ be a Lipschitz function defined on the torus $\mathbb{T}^1$. We are interested in the process $(\varphi \circ T^k)_k$. Let φ and ψ be two bounded functions on X. One has

$$\int_X \varphi \circ T.\psi \, d\mu = \int_X \varphi.P\psi \, d\mu,$$

where the operator P is given by

$$P\varphi(x) = \frac{1}{2}\left(\varphi\left(\frac{x}{2}\right) + \varphi\left(\frac{x+1}{2}\right) \right).$$

The iterates of this operator are

$$P^n\varphi(x) = \frac{1}{2^n} \sum_{k=0}^{2^n-1} \varphi\left(\frac{x+k}{2^n}\right).$$

For a Lipschitz function φ with Lipschitz constant L_φ one has

$$\left\| P^n\varphi - \int_0^1 \varphi \, d\mu \right\|_\infty \le \frac{L_\varphi}{2^n}.$$

For a centered Lipschitz function φ the sum $\sum_{k\ge 0} P^k\varphi$ is thus convergent.
  Besides one has

$$(TP\varphi)(x) = (P\varphi)(2x) = \frac{1}{2}\left(\varphi\left(\frac{2x}{2}\right) + \varphi\left(\frac{2x+1}{2}\right) \right)$$

$$= \frac{1}{2}\left(\varphi(x) + \varphi\left(x + \frac{1}{2}\right) \right),$$

and

$$(PT\varphi)(x) = \frac{1}{2}\left(T\varphi\left(\frac{x}{2}\right) + T\varphi\left(\frac{x+1}{2}\right) \right) = \frac{1}{2}(\varphi(x) + \varphi(x+1)) = \varphi(x).$$

Let $\mathscr{A}_0$ be the Borel σ-algebra of X, and $\mathscr{A}_n = T^{-n}\mathscr{A}_0$. If φ is $\mathscr{A}_0$-measurable, $T^n\varphi$ is $\mathscr{A}_n$-measurable. The atoms of $\mathscr{A}_n$ are the sets $\{x + \frac{k}{2^n} \ / \ k = 0, \ldots, 2^n - 1\}$ and we have

$$TP(\cdot) = \mathbb{E}(\cdot|\mathscr{A}_1) \quad T^n P^n(\cdot) = \mathbb{E}(\cdot|\mathscr{A}_n).$$

Let φ be a Lipschitz continuous function. One can write:

$$\varphi = \sum_{k \geq 0} P^k \varphi - \sum_{k \geq 1} P^k \varphi$$

$$= \sum_{k \geq 0} P^k \varphi - T \left(\sum_{k \geq 1} P^k \varphi \right) + T \left(\sum_{k \geq 1} P^k \varphi \right) - \sum_{k \geq 1} P^k \varphi.$$

Let f denote the function

$$f = \sum_{k \geq 0} P^k \varphi - T \left(\sum_{k \geq 1} P^k \varphi \right)$$

and h

$$h = -\sum_{k \geq 1} P^k \varphi,$$

then we have $\varphi = f + h - Th$ and

$$Pf = P \left(\sum_{k \geq 0} P^k \varphi \right) - PT \left(\sum_{k \geq 1} P^k \varphi \right) = \sum_{k \geq 1} P^k \varphi - \sum_{k \geq 1} P^k \varphi = 0$$

that is $\mathbb{E}(f|\mathscr{A}_1) = 0$. It also means that for $k > n$, $\mathbb{E}(T^n f|\mathscr{A}_k) = \mathbb{E}(f|\mathscr{A}_{k-n}) \circ T^n = 0$. In other words f generates a sequence of differences of a reversed martingale. Then f satisfies the CLT and so does φ.

1.2.3   The Gordin's Method

We consider a dynamical system $(X, \mathscr{A}, \mu, T)$.

Definition 1.4. We say that a function φ is cohomologous to a function generating a sequence of differences of martingale (*resp. reversed martingale*) if φ can be written $\varphi = f + h - Th$, with h measurable and f generating a sequence of differences of martingale (*resp. reversed martingale*) under the action of T.

The Gordin's method is based on the very simple fact that if φ and f are cohomologous, we have the relation

$$S_n \varphi = S_n f + h - T^n h,$$

between the ergodic sums, so that, after normalization, they have analogous behaviours.

The hilbertian characterization of notion of differences of martingale permits to obtain easily criteria for a function to be cohomologous to another one generating a sequence of differences of martingale.

For example we have the following theorem (cf. [25] page 145):

Theorem 1.4. *Let* $(X, \mathscr{A}, \mu, T)$ *be an invertible ergodic dynamical system,* $(\mathscr{A}_n)_{n \in \mathbb{Z}}$ *a filtration of* $\mathscr{A}$ *such that* $\mathscr{A}_n \supset \mathscr{A}_{n+1} = T^{-1}\mathscr{A}_n$ *(strictly) and* f *a function in* $L^2(\mu)$ *such that*

$$\sum_{n>0} \|\mathbb{E}(f|\mathscr{A}_n)\|_2 < \infty \text{ and } \sum_{n<0} \|f - \mathbb{E}(f|\mathscr{A}_n)\|_2 < \infty.$$

Then, either there exists a square integrable function h *such that* $f = h - Th$, *either the function* f *satisfies the CLT.*

Proof. The action of T on $L^2(\mathscr{B})$ by composition (which we still denote T) defines a unitary operator the adjoint of which is the composition by T^{-1}. To the filtration $(\mathscr{A}_n)$ one can associate a hilbertian filtration by considering the spaces $H_n = L^2(\mathscr{A}_n)$ of L^2-functions which are $\mathscr{A}_n$-measurable. Consider the restriction of T (the operator) to H_0. This has a sense because if f is in H_0 then Tf is $T^{-1}\mathscr{A}_0$-measurable, that is $\mathscr{A}_1$-measurable, otherwise said Tf is in $H_1 \subset H_0$. But T is not invertible on H_0 anymore. Let P the adjoint of the restriction of T to H_0. For f, g in H_0, one has by definition

$$\langle f, Tg \rangle = \langle Pf, g \rangle.$$

To say that a function $f \in H_0$ generates, under the action of T, a sequence of differences of a reversed martingale is to say that f is orthogonal to H_1. Indeed if this is the case, for every $k < l$, and $g \in H_0$, one has

$$\langle T^k f, T^l g \rangle = \langle f, T^{l-k} g \rangle = 0,$$

because $T^{l-k}g \in H_{l-k} \subset H_1$. This means that $\mathbb{E}(T^k f | \mathscr{A}_l) = 0$, which means that $(T^k f)_k$ is a sequence of differences of a reversed martingale relatively to the filtration $(\mathscr{A}_l)$. Suppose that for some $f \in H_0$ the series $\sum_{k \geq 0} P^k f$ does converge. Then write

$$f = \sum_{k \geq 0} P^k f - \sum_{k \geq 1} P^k f$$

$$= \sum_{k \geq 0} P^k f - T\left(\sum_{k \geq 1} P^k f\right) + T\left(\sum_{k \geq 1} P^k f\right) - \sum_{k \geq 1} P^k f.$$

The last two terms of the last expression define a coboundary. Let us show the sum
of the first two generate a sequence of differences of a reversed martingale. As noted
above we just have to show that $\varphi_+ = \sum_{k \geq 0} P^k f - T \left(\sum_{k \geq 1} P^k f \right)$ is orthogonal
to $H_1 = TH_0$. For $g \in H_0$ we compute:

$$\left\langle \sum_{k \geq 0} P^k f - T \left(\sum_{k \geq 1} P^k f \right), Tg \right\rangle = \left\langle \sum_{k \geq 0} P^k f, Tg \right\rangle - \left\langle T \left(\sum_{k \geq 1} P^k f \right), Tg \right\rangle$$

$$= \left\langle P \sum_{k \geq 0} P^k f, g \right\rangle - \left\langle \sum_{k \geq 1} P^k f, g \right\rangle$$

$$= \left\langle \sum_{k \geq 1} P^k f, g \right\rangle - \left\langle \sum_{k \geq 1} P^k f, g \right\rangle = 0,$$

the second equality is due to the definition of P and the fact that T is an isometry
of H_0.

We can do something similar on the orthogonal of H_0. Let Q be the adjoint of
the restriction of T^{-1} to $H_0^\perp$: for f, g in $H_0^\perp$, one has

$$\langle f, T^{-1}g \rangle = \langle Qf, g \rangle.$$

Suppose that for some $f \in H_0^\perp$ the series $\sum_{k \geq 0} Q^k f$ does converge. Then write

$$f = \sum_{k \geq 0} Q^k f - \sum_{k \geq 1} Q^k f$$

$$= \sum_{k \geq 0} Q^k f - T^{-1} \left(\sum_{k \geq 1} Q^k f \right) + T^{-1} \left(\sum_{k \geq 1} Q^k f \right) - \sum_{k \geq 1} Q^k f.$$

We have a coboundary part and we want to show that

$$\varphi_- = \sum_{k \geq 0} Q^k f - T^{-1} \left(\sum_{k \geq 1} Q^k f \right)$$

generates a sequences of differences of a reversed martingale, that is in $H_0^\perp$ and
orthogonal to $H_{-1}^\perp$. For $g \in H_0^\perp$ we compute:

$$\langle \sum_{k \geq 0} Q^k f - T^{-1}\left(\sum_{k \geq 1} Q^k f\right), T^{-1}g\rangle = \langle \sum_{k \geq 0} Q^k f, T^{-1}g\rangle$$

$$-\langle T^{-1}\left(\sum_{k \geq 1} Q^k f\right), T^{-1}g\rangle$$

$$= \langle Q \sum_{k \geq 0} Q^k f, g\rangle - \langle \left(\sum_{k \geq 1} Q^k f\right), g\rangle$$

$$= 0.$$

Now we give relations between P, Q, T, T^{-1} and the conditional expectations. For f, g in H_0, one has

$$\langle f, PTg\rangle = \langle Tf, Tg\rangle = \langle f, g\rangle,$$

so that $PT = Id_{H_0}$. On the other hand, one has $TPTP = TP$ and

$$\langle f - TPf, Tg\rangle = \langle f, Tg\rangle - \langle TPf, Tg\rangle = \langle Pf, g\rangle - \langle Pf, g\rangle = 0,$$

so that TP is the orthogonal projection of H_0 on H_1. In other words, for $f \in H_0$

$$TPf = \mathbb{E}(f|\mathscr{A}_1),$$

and similarly

$$T^k P^k f = \mathbb{E}(f|\mathscr{A}_k).$$

For f, g in $H_0^\perp$, one has

$$\langle f, QT^{-1}g\rangle = \langle T^{-1}f, T^{-1}g\rangle = \langle f, g\rangle,$$

so that $QT^{-1} = Id_{H_0^\perp}$. One has $T^{-1}QT^{-1}Q = T^{-1}Q$ and

$$\langle f - T^{-1}Qf, T^{-1}g\rangle = \langle f, T^{-1}g\rangle - \langle T^{-1}Qf, T^{-1}g\rangle = \langle Qf, g\rangle - \langle Qf, g\rangle = 0,$$

so that $T^{-1}Q$ is the orthogonal projection of $H_0^\perp$ on $H_{-1}^\perp$. In other words, as the elements $f \in H_0^\perp$ are the functions $g - \mathbb{E}(g|\mathscr{A}_0)$,

$$T^{-1}Q(g - \mathbb{E}(g|\mathscr{A}_0)) = g - \mathbb{E}(g|\mathscr{A}_{-1}),$$

and similarly

$$T^{-k}Q^k(g - \mathbb{E}(g|\mathscr{A}_0)) = g - \mathbb{E}(g|\mathscr{A}_{-k}).$$

In particular for every L^2-function f one has

$$\| P^k \mathbb{E}(f|\mathscr{A}_0)\|_2 = \|\mathbb{E}(f|\mathscr{A}_k)\|_2 \text{ and } \| Q^k(f - \mathbb{E}(f|\mathscr{A}_0))\|_2 = \| f - \mathbb{E}(f|\mathscr{A}_{-k})\|_2.$$

This shows that if the two series

$$\sum_{n>0} \|\mathbb{E}(f|\mathscr{A}_n)\|_2 \text{ and } \sum_{n<0} \| f - \mathbb{E}(f|\mathscr{A}_n)\|_2$$

converge then the two series

$$\sum_{n>0} \| P^n \mathbb{E}(f|\mathscr{A}_0)\|_2 \text{ and } \sum_{n>0} \| Q^n(f - \mathbb{E}(f|\mathscr{A}_0))\|_2$$

also, and both $\mathbb{E}(f|\mathscr{A}_0)$ and $f - \mathbb{E}(f|\mathscr{A}_0)$ are cohomologous to functions generating sequences of differences of reversed martingales. It implies that this is also the case for $f = f - \mathbb{E}(f|\mathscr{A}_0) + \mathbb{E}(f|\mathscr{A}_0)$.

The convergencies of the series imply that one can write $f = g + h - Th$ with g generating a sequence of differences of a reversed martingale. If g is not 0 the asymptotic variance

$$\sigma^2 = \int f^2 \, d\mu + 2 \sum_{k=1}^{\infty} \langle f, T^k f \rangle = \int g^2 \, d\mu$$

is positive and f satisfies the CLT (because g does by Theorem 1.3). $\qquad \square$

Let us explain how to use the preceding theorem implies limit theorems for quasi-hyperbolic dynamical systems. The same result holds with an increasing filtration (but one needs to exchange the signs $n > 0$ and $n < 0$ in the series and one obtains a sequence of differences of martingale).

Suppose that X is a manifold and that at each point one can define an unstable manifold for T. Suppose moreover that one can construct a filtration $(\mathscr{A}_n)$ such that, for every n, the atoms de $\mathscr{A}_n$ are pieces of unstable leaves. Then the atoms of $\mathscr{A}_n$, images by T^{-n} of the atoms of $\mathscr{A}_0$, are very small when n goes to $+\infty$ and whirl in X when n goes to $-\infty$. Consider a C^∞ function f on X. The value of the conditional expectation $\mathbb{E}(f|\mathscr{A}_n)$ at a point x is the mean value of f on the atom of $\mathscr{A}_n$ containing x. For n very large $\mathbb{E}(f|\mathscr{A}_n)(x)$ is very near $f(x)$: the convergence of the second series will be easy to establish. For n very negative $\mathbb{E}(f|\mathscr{A}_n)(x)$ is the integral of f on a large piece of unstable leaf passing at x: to show the convergence of the first series, one need to dispose of a result of equidistribution of the unstable leaves of T in X. This point is more difficult.

We will do some computations with conditional expectations. Let $(X, \mathscr{A}, \mu, T)$ be an invertible ergodic dynamical system, $(\mathscr{A}_n)$ a filtration of $\mathscr{A}$ such that $\mathscr{A}_n \subset \mathscr{A}_{n+1} = T^{-1}\mathscr{A}_n$. Let g a function measurable with respect to $\mathscr{A}_0$. Then $g \circ T^k$ is measurable with respect to $\mathscr{A}_k$. Now by definition

$$\langle \mathbb{E}\left(T^l f | \mathscr{A}_k\right), T^k g \rangle = \int \mathbb{E}\left(f \circ T^l | \mathscr{A}_k\right) g \circ T^k \, d\mu = \int f \circ T^l \cdot g \circ T^k \, d\mu.$$

Then one can use the invariance of μ under the action of T to get

$$\int f \circ T^l \cdot g \circ T^k \, d\mu = \int f \circ T^{l-k} \cdot g \, d\mu,$$

and

$$\int f \circ T^{l-k} \cdot g \, d\mu = \int \mathbb{E}\left(f \circ T^{l-k} | \mathscr{A}_0\right) g \, d\mu = \int \mathbb{E}\left(f \circ T^{l-k} | \mathscr{A}_0\right) \circ T^k \, g \circ T^k \, d\mu.$$

We thus have obtained

$$\mathbb{E}\left(f \circ T^l | \mathscr{A}_k\right) = \mathbb{E}\left(f \circ T^{l-k} | \mathscr{A}_0\right) \circ T^k,$$

a relation we will use several times below.

1.2.4   Example 2: The Cat Map

We begin by a simple example on which the method works without the technical difficulties that will arise later. We also give on this example more geometrical descriptions of the quantities defined above in the proof of Theorem 1.4. We change a little the construction to get a sequence of differences of martingales (rather than reversed martingales).

Let $A = \begin{pmatrix} 2 & 1 \\ 1 & 1 \end{pmatrix}$ and consider the map

$$T : \mathbb{T}^2 \longrightarrow \mathbb{T}^2$$

$$x \longmapsto Ax.$$

This, together with the Lebesgue measure m (invariant by T) defines the dynamical system $(\mathbb{T}^2, T, m)$.

A convenient filtration is obtained here by considering the σ-algebra $\mathscr{A}_0$ the atoms of which are the unstable sections of the three rectangles drawn below. These rectangles define a Markov partition for T: the images by T of the unstable section of the rectangles are finite unions of such sections (see Fig. 1.1); hence the inclusion $\mathscr{A}_k = T^{-k} \mathscr{A}_0 \subset \mathscr{A}_{k+1} = T^{-k-1} \mathscr{A}_0$.

First, we will write series regardless of their convergence.

Let us define the operator P by

$$P(f) = \mathbb{E}(f | \mathscr{A}_{-1}) \circ T = \mathbb{E}(Tf | \mathscr{A}_0).$$

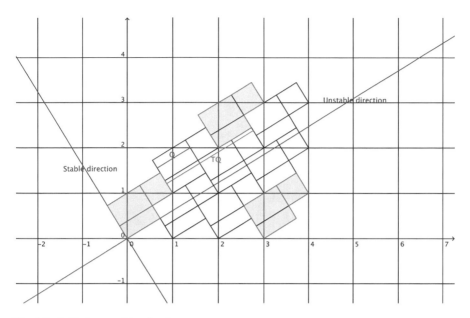

Fig. 1.1 A Markov partition for the cat map

The iterates of P are given by

$$P^2(f) = P\mathbb{E}(Tf|\mathscr{A}_0) = \mathbb{E}(T\mathbb{E}(Tf|\mathscr{A}_0)|\mathscr{A}_0)$$
$$= \mathbb{E}(\mathbb{E}(T^2 f|\mathscr{A}_1)|\mathscr{A}_0) = \mathbb{E}(T^2 f|\mathscr{A}_0)$$

and similarly

$$P^k(f) = \mathbb{E}(T^k f|\mathscr{A}_0).$$

Remark that P is an operator of the space $L^2(\mathscr{A}_0)$

$$PT^{-1}\mathbb{E}(f|\mathscr{A}_0) = P\left(\mathbb{E}(f|\mathscr{A}_0) \circ T^{-1}\right) = \mathbb{E}\left(\mathbb{E}(f|\mathscr{A}_0) \circ T^{-1} \circ T|\mathscr{A}_0\right) = \mathbb{E}(f|\mathscr{A}_0),$$

that is PT^{-1} is the identity on the space $L^2(\mathscr{A}_0)$. As in the case of the angle doubling we write

$$\mathbb{E}(f|\mathscr{A}_0) = \sum_{k\geq 0} P^k \mathbb{E}(f|\mathscr{A}_0) - \sum_{k\geq 1} P^k \mathbb{E}(f|\mathscr{A}_0)$$

$$= \sum_{k\geq 0} P^k \mathbb{E}(f|\mathscr{A}_0) - T^{-1} \sum_{k\geq 1} P^k \mathbb{E}(f|\mathscr{A}_0)$$

$$+ T^{-1} \sum_{k\geq 1} P^k \mathbb{E}(f|\mathscr{A}_0) - \sum_{k\geq 1} P^k \mathbb{E}(f|\mathscr{A}_0).$$

Let us denote φ_+ the function

$$\varphi_+ = \sum_{k \geq 0} P^k \mathbb{E}(f|\mathscr{A}_0) - T^{-1} \sum_{k \geq 1} P^k \mathbb{E}(f|\mathscr{A}_0)$$

and ψ_+

$$\psi_+ = T^{-1} \sum_{k \geq 1} P^k \mathbb{E}(f|\mathscr{A}_0).$$

Then we have $\mathbb{E}(f|\mathscr{A}_0) = \varphi_+ + \psi_+ - T\psi_+$ and $P\varphi_+ = 0$ because

$$P\varphi_+ = P \sum_{k \geq 0} P^k \mathbb{E}(f|\mathscr{A}_0) - PT^{-1} \sum_{k \geq 1} P^k \mathbb{E}(f|\mathscr{A}_0)$$

$$= \sum_{k \geq 1} P^k \mathbb{E}(f|\mathscr{A}_0) - \sum_{k \geq 1} P^k \mathbb{E}(f|\mathscr{A}_0)$$

$$= 0$$

(because $PT^{-1} = Id$ on $L^2(\mathscr{A}_0)$). The function φ_+ is in $L^2(\mathscr{A}_0)$ and orthogonal to $L^2(\mathscr{A}_{-1})$: it generates a sequence of differences of a martingale.

Now, we will do something similar with $f - \mathbb{E}(f|\mathscr{A}_0)$.

Consider the operator Q defined by

$$Qg = (g - \mathbb{E}(g|\mathscr{A}_1)) \circ T^{-1} = g \circ T^{-1} - \mathbb{E}(T^{-1}g|\mathscr{A}_0).$$

The second expression of Qg shows that the image of Q is included in the orthogonal of $L^2(\mathscr{A}_0)$. Moreover, $Qg = 0$ if g is in $L^2(\mathscr{A}_0)$ and, if g is orthogonal to $L^2(\mathscr{A}_0)$, that is, if $g = h - \mathbb{E}(h|\mathscr{A}_0)$ for some h, we have

$$QTg = g - \mathbb{E}(g|\mathscr{A}_0) = h - \mathbb{E}(h|\mathscr{A}_0) - \mathbb{E}(h|\mathscr{A}_0) + \mathbb{E}(h|\mathscr{A}_0) = g.$$

Let us write the iterates of Q; first the second one

$$Q^2 g = Q\left(g \circ T^{-1} - \mathbb{E}(T^{-1}g|\mathscr{A}_0)\right)$$

$$= Q\left(g \circ T^{-1}\right)$$

$$= g \circ T^{-2} - \mathbb{E}(T^{-2}g|\mathscr{A}_0)$$

$$= g \circ T^{-2} - \mathbb{E}(g|\mathscr{A}_2) \circ T^{-2}$$

and similarly:

$$Q^k g = g \circ T^{-k} - \mathbb{E}(g|\mathscr{A}_k) \circ T^{-k}$$

and

$$Q^k(f - \mathbb{E}(f|\mathscr{A}_0)) = T^{-k}f - \mathbb{E}(f|\mathscr{A}_k) \circ T^{-k}.$$

Now we can write

$$f - \mathbb{E}(f|\mathscr{A}_0) = \sum_{k \geq 0} Q^k(f - \mathbb{E}(f|\mathscr{A}_0)) - \sum_{k \geq 1} Q^k(f - \mathbb{E}(f|\mathscr{A}_0))$$

$$= \sum_{k \geq 0} Q^k(f - \mathbb{E}(f|\mathscr{A}_0)) - T\left(\sum_{k \geq 1} Q^k(f - \mathbb{E}(f|\mathscr{A}_0))\right)$$

$$+ T\left(\sum_{k \geq 1} Q^k(f - \mathbb{E}(f|\mathscr{A}_0))\right) - \sum_{k \geq 1} Q^k(f - \mathbb{E}(f|\mathscr{A}_0)).$$

Let us denote φ_- the function

$$\varphi_- = \sum_{k \geq 0} Q^k(f - \mathbb{E}(f|\mathscr{A}_0)) - T\left(\sum_{k \geq 1} Q^k(f - \mathbb{E}(f|\mathscr{A}_0))\right)$$

and ψ_-

$$\psi_- = -\sum_{k \geq 1} Q^k(f - \mathbb{E}(f|\mathscr{A}_0)).$$

Then $f - \mathbb{E}(f|\mathscr{A}_0) = \varphi_- + \psi_- - T\psi_-$ and $Q\varphi_- = 0$ because

$$Q\varphi_- = Q \sum_{k \geq 0} Q^k(f - \mathbb{E}(f|\mathscr{A}_0)) - QT \sum_{k \geq 1} Q^k(f - \mathbb{E}(f|\mathscr{A}_0))$$

$$= \sum_{k \geq 1} Q^k(f - \mathbb{E}(f|\mathscr{A}_0)) - \sum_{k \geq 1} Q^k(f - \mathbb{E}(f|\mathscr{A}_0))$$

$$= 0$$

as we have seen that $QT = Id$ on the orthogonal of $L_2(\mathscr{A}_0)$. This means that φ_- is in $L_2(\mathscr{A}_1)$ and orthogonal to $L_2(\mathscr{A}_0)$: thus generates a sequence of differences of martingales.

So we have obtained an equality

$$f = \varphi_+ + \varphi_- + \psi - T\psi$$

where φ is in $L_2(\mathscr{A}_0)$ orthogonal to $L_2(\mathscr{A}_{-1})$, φ_- is in $L_2(\mathscr{A}_1)$ orthogonal to $L_2(\mathscr{A}_0)$. Modifying a little bit our expression we obtain

$$f = \varphi + \chi - T\chi$$

where φ is in $L_2(\mathscr{A}_0)$ orthogonal to $L_2(\mathscr{A}_{-1})$. In other words f is cohomologous to the function φ that generates a sequence of differences of a martingale *if the series considered do converge*. The question is thus: do the series

$$\sum_{k \geq 0} P^k \mathbb{E}(f|\mathscr{A}_0) \text{ and } \sum_{k \geq 0} Q^k(f - \mathbb{E}(f|\mathscr{A}_0))$$

converge? In view of the expressions of the iterates of P and Q this question becomes: do the series

$$\sum_{k \geq 0} \mathbb{E}(f|\mathscr{A}_{-k}) \circ T^k \text{ and } \sum_{k \geq 0} (f - \mathbb{E}(f|\mathscr{A}_k)) \circ T^{-k}$$

converge? This is the case if the series

$$\sum_{k \geq 0} \|E(f|\mathscr{A}_{-k})\|_\infty < \infty \text{ and } \sum_{k \geq 0} \|f - E(f|\mathscr{A}_k)\|_\infty < \infty$$

converge.

Let us show that in our example these two series converge when f is a C^∞ function. Let us write f as the sum of its Fourier series:

$$f = \sum_{p \in \mathbb{Z}^2} c_p e_p$$

where e_p denotes the exponential function $e_p(\cdot) = \exp(2i\pi\langle p, \cdot\rangle)$ and c_p is the Fourier coefficient

$$c_p = \int_{\mathbb{T}^2} f(x) \exp(-2i\pi\langle p, x\rangle) dx.$$

The atoms of $\mathscr{A}_k$ are line segment of the form

$$\mathscr{A}_k(x) = \{x + tv \, / \, t \in [l_{k,x}, u_{k,x}]\}$$

where v is the vector $(\frac{\sqrt{5}-1}{2}, 1)$ (eigenvector associated to the eigenvalue $\lambda = \frac{3+\sqrt{5}}{2}$). The value of $\mathbb{E}(f|\mathscr{A}_k)(x)$ is almost surely defined by

$$\mathbb{E}(f|\mathscr{A}_k)(x) = \frac{1}{(u_{k,x} - l_{k,x})} \int_{l_{k,x}}^{u_{k,x}} f(x + tv) dt.$$

For a fixed k, these segments may have two different lengths: the numbers $u_{k,x} - l_{k,x}$ belongs to a pair $\{a\lambda^{-k}, b\lambda^{-k}\}$ (where a and b are the two numbers necessary to the description of the two length of the unstable side of the rectangles of the Markov partition). For positive k, the difference between $\mathbb{E}(f \,|\, \mathscr{A}_k)(x)$ and $f(x)$ is exponentially small in k:

$$
|f(x) - \mathbb{E}(f \,|\, \mathscr{A}_k)(x)| = \left| f(x) - \frac{1}{(u_{k,x} - l_{k,x})} \int_{l_{k,x}}^{u_{k,x}} f(x + tv)dt \right|
$$

$$
= \frac{1}{(u_{k,x} - l_{k,x})} \left| \int_{l_{k,x}}^{u_{k,x}} (f(x) - f(x + tv))\, dt \right|
$$

$$
\leq \frac{1}{(u_{k,x} - l_{k,x})} \int_{l_{k,x}}^{u_{k,x}} |f(x) - f(x + tv)|\, dt
$$

$$
\leq \frac{1}{(u_{k,x} - l_{k,x})} \int_{l_{k,x}}^{u_{k,x}} \|\nabla f\|_\infty \|v\| |t|\, dt
$$

$$
\leq C \|\nabla f\|_\infty \lambda^{-k}.
$$

For negative k, the following computations show the equidistribution of unstable leaves of the cat map:

$$
|\mathbb{E}(f \,|\, \mathscr{A}_k)(x)| = \left| \frac{1}{(u_{k,x} - l_{k,x})} \int_{l_{k,x}}^{u_{k,x}} f(x + tv)dt \right|
$$

$$
= \frac{1}{(u_{k,x} - l_{k,x})} \left| \int_{l_{k,x}}^{u_{k,x}} \sum_{p \in \mathbb{Z}^2} c_p \exp(2i\pi \langle x + tv, p \rangle)dt \right|
$$

$$
= \left| \sum_{p \in \mathbb{Z}^2} c_p \exp(2i\pi \langle x, p \rangle) \frac{1}{(u_{k,x} - l_{k,x})} \int_{l_{k,x}}^{u_{k,x}} \exp(2i\pi \langle tv, p \rangle)dt \right|
$$

$$
\leq \sum_{p \in \mathbb{Z}^2} \left| c_p \exp(2i\pi \langle x, p \rangle) \frac{1}{(u_{k,x} - l_{k,x})} \int_{l_{k,x}}^{u_{k,x}} \exp(2i\pi \langle tv, p \rangle)dt \right|
$$

$$
\leq \sum_{p \in \mathbb{Z}^2} |c_p| \frac{1}{(u_{k,x} - l_{k,x})} \left| \int_{l_{k,x}}^{u_{k,x}} \exp(2i\pi \langle tv, p \rangle)dt \right|
$$

$$
\leq \sum_{p \in \mathbb{Z}^2} |c_p| \frac{1}{(u_{k,x} - l_{k,x})} \left| \int_{l_{k,x}}^{u_{k,x}} \exp(2i\pi t(\frac{\sqrt{5}-1}{2} p_1 + p_2))dt \right|
$$

$$
\leq \sum_{p \in \mathbb{Z}^2} |c_p| \frac{1}{(u_{k,x} - l_{k,x})} \frac{2}{|2i\pi(\frac{\sqrt{5}-1}{2} p_1 + p_2)|}
$$

$$
\leq C \lambda^k \sum_{p \in \mathbb{Z}^2} |c_p| \frac{1}{|\frac{\sqrt{5}-1}{2} p_1 + p_2|}.
$$

But, $\frac{\sqrt{5}-1}{2}$ is quadratic, so that $|\frac{\sqrt{5}-1}{2}p_1 + p_2| \geq c|p_1|^{-1}$. Thus we have

$$\left|\mathbb{E}(f|\mathscr{A}_k)(x)\right| \leq C\lambda^k \sum_{p\in\mathbb{Z}^2} |p_1 c_p|.$$

If f is C^∞ the series $\sum_{p\in\mathbb{Z}^2} |p_1 c_p|$ converges. The two series

$$\sum_{k\geq 0} \|E(f|\mathscr{A}_{-k})\|_\infty < \infty \quad \text{and} \quad \sum_{k\geq 0} \|f - E(f|\mathscr{A}_k)\|_\infty < \infty$$

are thus convergent in this case.

1.3 Other Limit Theorems and Construction of Adequate Filtrations

1.3.1 Some Other Limit Theorems

1.3.1.1 The Donsker Invariance Principle

We are interested now in another limit theorem. This is also a convergence in distribution but for probabilities defined on the space of continuous functions on $[0, 1]$. The Donsker invariance principle states the convergence toward the Wiener measure of some processes with values in $(\mathscr{C}([0, 1]), \|\,\|_\infty)$.

Theorem 1.5. *There exists on $\mathscr{C}([0, 1])$ a unique measure $\mathbb{P}$ for which the following properties hold:*

  (i) *For every t, the law of the random variable $W_t : f \mapsto f(t)$ is the centered gaussian law with variance t.*
 (ii) *For every t, s with $s < t$, the random variables $W_t - W_s$ and W_s are independent.*
(iii) *For every t, s with $s < t$, the random variable $W_t - W_s$ has the same distribution as W_{t-s}.*

The probability which is defined in the preceding theorem is the celebrated Wiener measure.

Proposition 1.1. *A probability on $\mathscr{C}([0, 1])$ is uniquely determined by its finite dimensional projections that is the distributions of $(Y_{t_1}, Y_{t_2}, \ldots, Y_{t_k})$ for every $k \in \mathbb{N}$ and $(t_1, t_2, \ldots, t_k) \in [0, 1]^k$.*

It is well explained in [4] how to get the convergence towards the Wiener measure (and Billingsley treat the case of the sequences of differences of a martingale; this is also done in the short paper [9]). One can proceed in two steps: first get the convergence of finite dimensional distributions, second prove that the sequence considered is tight.

A sequence of probabilities (μ_n) seen as a Radon measure on a polish space X is said to be tight if for every $\epsilon > 0$ one can find a compact subset K of X such that, for every n, one has $\mu_n(K) > 1 - \epsilon$.

Arzela-Ascoli theorem characterizes the relatively compact subsets of $\mathscr{C}([0, 1])$: these are the sets $E \subset \mathscr{C}([0, 1])$ such that, for every $x \in [0, 1]$, $\sup_{f \in E} |f(x)| < \infty$ and E is equicontinuous, that is

$$\forall \epsilon > 0 \; \exists \alpha > 0 \; \forall x, y \; (|x - y| < \alpha \Rightarrow |f(x) - f(y)| < \epsilon).$$

Let (ξ_n) be a sequence of random variables with values in $\mathscr{C}([0, 1])$. The sequence of the image probabilities on $\mathscr{C}([0, 1])$ is tight if (we just write that the probabilities of large compact sets of $\mathscr{C}([0, 1])$ are uniformly near 1): for every $x \in [0, 1]$,

$$\lim_{M \to \infty} \sup_n \mathbb{P}(|\xi_n(x)| > M) = 0$$

and, for every $\epsilon > 0$,

$$\lim_{\alpha \to 0} \sup_n \mathbb{P}(\sup_{|x-y|<\alpha} |\xi_n(x) - \xi_n(y)| > \epsilon) = 0.$$

The tightness of a normalized sequence defined by a bounded stationary ergodic martingale is a consequence of the following maximal inequality.

Theorem 1.6 ("Doob inequality"[2]). *Let $(\mathscr{F}_k)_{k \geq 0}$ be a filtration on probability space $(\Omega, \mathscr{F}, \mathbb{P})$ and let $(S_k)_{k \geq 0}$ be a martingale with respect to the filtration $(\mathscr{F}_k)_{k \geq 0}$. Then we have*

$$\mathbb{P}\left(\sup_{0 \leq k \leq n} |S_k^4| \geq \alpha\right) \leq \frac{\mathbb{E}\left(S_n^4\right)}{\alpha}.$$

Proof. Let A_j be the set

$$A_j = \sup_{0 \leq k \leq j} S_k^4 < \alpha.$$

The sequence of sets obtained is decreasing and their intersection is A_n the complementary set of which is $B = \sup_{0 \leq k \leq n} S_k^4 \geq \alpha$. Let us consider the sum

---

[2]This inequality is true for non negative submartingales.

$$U = \sum_{j=0}^{n-1} \mathbf{1}_{A_j} (S_{j+1}^4 - S_j^4) + S_0^4.$$

If $x \in B$ then there exists j such that $x \notin A_j$. The variable U is then equal at x to $S_{j+1}^4(x)$ where j satisfies $\sup_{0 \le k \le j} S_k^4(x) < \alpha$, $\sup_{0 \le k \le j+1} S_k^4 \ge \alpha$, so that $U(x) = S_{j+1}^4(x) \ge \alpha$; one has $\alpha \mathbf{1}_B \le U$ so that

$$\mathbb{P}(B) \le \mathbb{E}(U)/\alpha,$$

and we just have to check that $\mathbb{E}(U) \le \mathbb{E}(S_n^4)$.

$$\mathbb{E}(U) = \mathbb{E}\left(\sum_{j=0}^{n-1} \mathbf{1}_{A_j} (S_{j+1}^4 - S_j^4) \right) + \mathbb{E}(S_0^4)$$

$$= \mathbb{E}\left(\sum_{j=0}^{n-1} (S_{j+1}^4 - S_j^4) \right) - \mathbb{E}\left(\sum_{j=0}^{n-1} \mathbf{1}_{{}^cA_j} (S_{j+1}^4 - S_j^4) \right) + \mathbb{E}(S_0^4)$$

$$= \mathbb{E}(S_n^4) - \sum_{j=0}^{n-1} \mathbb{E}\left(\mathbf{1}_{{}^cA_j} (S_{j+1}^4 - S_j^4) \right)$$

$$= \mathbb{E}(S_n^4) - \sum_{j=0}^{n-1} \mathbb{E}\left(\mathbf{1}_{{}^cA_j} (\mathbb{E}(S_{j+1}^4 | \mathscr{F}_j) - S_j^4) \right)$$

But the Jensen inequality gives $\mathbb{E}(S_{j+1}^4 | \mathscr{F}_j) \ge \mathbb{E}(S_{j+1} | \mathscr{F}_j)^4 = S_j^4$ (because $t \mapsto t^4$ is convex). Hence the result. $\square$

Let (Y_k) be a sequence of random variables. Let us define a random element of $\mathscr{C}([0, 1])$:

$$\xi_n(t) = \frac{1}{\sigma \sqrt{n}} \left(S_k + n(t - \frac{k}{n}) Y_{k+1} \right),$$

for $t \in [\frac{k}{n}, \frac{k+1}{n}], k = 0, \ldots, n - 1$.

Theorem 1.7. *Let Y_k a sequence of differences of martingale (or of reversed martingale) with variance 1. One has the convergence (in distribution)*

$$\xi_n \to^{\mathscr{L}} W.$$

Proof. We begin by the convergence of finite dimensional distributions. Let l be a positive integer, $t_1, \ldots, t_l, l$ numbers in $[0, 1]$. We want to show that

$$(\xi_n(t_1), \ldots, \xi_n(t_l)) \to^{\mathscr{L}} (W_{t_1}, \ldots, W_{t_l}).$$

It is equivalent to the following convergence (we just apply a linear transformation) which gives simpler computations:

$$(\xi_n(t_1), \xi_n(t_2) - \xi_n(t_1), \ldots, \xi_n(t_l) - \xi_n(t_{l-1})) \to^{\mathscr{L}} (W_{t_1}, W_{t_2} - W_{t_1}, \ldots, W_{t_l} - W_{t_{l-1}}).$$

Making use of the theorem of Levy it reduces to show that, for all u_j,

$$\mathbb{E}\left(\exp\left(i \sum_{j=1}^{l} u_j(\xi_n(t_j) - \xi_n(t_{j-1}))\right)\right) \to \prod_{j=1}^{l} \exp(-u_j^2(t_j - t_{j-1})/2).$$

One more time, we consider just the bounded case. This implies that

$$\left| \xi_n(t_j) - \xi_n(t_{j-1}) - \frac{1}{\sqrt{n}} \sum_{k=[nt_{j-1}]}^{[nt_j]} Y_k \right| \le \frac{C}{\sqrt{n}}.$$

We now use again the expression given by Lemma 1.1:

$$\mathbb{E}\left(\exp(i \sum_{j=1}^{l} \frac{u_j}{\sqrt{n}} \sum_{k=[nt_{j-1}]}^{[nt_j]} Y_k)\right)$$

$$= \mathbb{E}\left(\prod_{j=1}^{l} \prod_{k=[nt_{j-1}]}^{[nt_j]} \exp(i \frac{u_j Y_k}{\sqrt{n}})\right)$$

$$= \mathbb{E}\left(\prod_{j=1}^{l} \prod_{k=[nt_{j-1}]}^{[nt_j]} (1 + i \frac{u_j Y_k}{\sqrt{n}}) \exp(-\frac{u_j^2 Y_k^2}{n} + \varphi(\frac{u_j Y_k}{\sqrt{n}}))\right)$$

$$= \mathbb{E}\left(\prod_{j=1}^{l} \prod_{k=[nt_{j-1}]}^{[nt_j]} (1 + i \frac{u_j Y_k}{\sqrt{n}}) \left(\prod_{j=1}^{l}(\exp(-\frac{u_j^2}{n} \sum_{k=[nt_{j-1}]}^{[nt_j]} Y_k^2 + \varphi(\frac{u_j Y_k}{\sqrt{n}}))\right)\right).$$

Birkhoff theorem implies that

$$-\frac{u_j^2}{n} \sum_{k=[nt_{j-1}]}^{[nt_j]} Y_k^2 \to -u_j^2(t_j - t_{j-1})\mathbb{E}(Y_1^2)/2 = -u_j^2(t_j - t_{j-1})/2.$$

As the Y_k are bounded and $\varphi(t) \le Ct^3$, the Lebesgue theorem implies that the difference between

$$\mathbb{E}\left(\prod_{j=1}^{l}\prod_{k=[nt_{j-1}]}^{[nt_j]}(1+i\frac{u_j Y_k}{\sqrt{n}})\left(\prod_{j=1}^{l}(\exp(-\frac{u_j^2}{n}\sum_{k=[nt_{j-1}]}^{[nt_j]}Y_k^2+\varphi(\frac{u_j Y_k}{\sqrt{n}}))\right)\right)$$

and

$$\mathbb{E}\left(\prod_{j=1}^{l}\prod_{k=[nt_{j-1}]}^{[nt_j]}(1+i\frac{u_j Y_k}{\sqrt{n}})\left(\prod_{j=1}^{l}\exp(-u_j^2(t_j-t_{j-1})/2)\right)\right)$$

tends to 0. At last the martingale property insures that

$$\mathbb{E}\left(\prod_{j=1}^{l}\prod_{k=[nt_{j-1}]}^{[nt_j]}(1+i\frac{u_j Y_k}{\sqrt{n}})\right)=1.$$

The convergence of finite dimensional distributions is now established. We will use the elementary following fact (we are in the bounded case): there exists $C > 0$, such that

$$\mathbb{E}(S_n^4)\leq Cn^2.$$

This is an easy consequence of the martingale property (develop the fourth power of the sum S_n and take the expectation; the only terms lefts are the n terms $\mathbb{E}(Y_i^4)$ and the $n(n-1)/2$ terms $\mathbb{E}(Y_i^2 Y_j^2)$). For n large enough, one has

$$\mathbb{P}(|\xi_n(t)| > M)\leq \mathbb{P}(\|\frac{1}{\sqrt{n}}\sum_{k=0}^{[nt]}Y_k\|\geq M-1)$$

$$\leq \mathbb{P}((\sum_{k=0}^{[nt]}Y_k)^4\geq (M-1)^4 n^2)\leq \frac{\mathbb{E}(S_{[nt]}^4)}{(M-1)^4 n^2}\leq Ct^4(M-1)^{-4},$$

so that $\lim_{M\to\infty}\mathbb{P}(|\xi_n(t)| > M) = 0$. At last it remains to prove that, for every $\epsilon > 0$,

$$\lim_{\alpha\to 0}\sup_n \mathbb{P}(\sup_{|x-y|<\alpha}|\xi_n(x)-\xi_n(y)| > \epsilon) = 0.$$

Let us fix ϵ and α. If $\sup_{|x-y|<\alpha}|\xi_n(x)-\xi_n(y)| > \epsilon$ then there exists an integer $k\leq 1/\alpha$ such that $\sup_{|x-k\alpha|<\alpha}|\xi_n(x)-\xi_n(k\alpha)| > \epsilon/2$. As

$$\mathbb{P}(\sup_{|x-k\alpha|<\alpha}|\xi_n(x)-\xi_n(k\alpha)|>\epsilon/2)\leq\mathbb{P}(\sup_{0\leq\ell\leq\alpha(n+1)}\left|\frac{1}{\sqrt{n}}\sum_{j=[n(k-1)\alpha]}^{[n(k-1)\alpha]+\ell}Y_j\right|\geq\epsilon/2)$$

$$+\mathbb{P}(\sup_{0\leq\ell\leq\alpha(n+1)}\left|\frac{1}{\sqrt{n}}\sum_{j=[nk\alpha]}^{[nk\alpha]+\ell}Y_j\right|\geq\epsilon/2),$$

the Doob inequality implies that

$$\mathbb{P}(\sup_{0\leq\ell\leq\alpha(n+1)}\left|\frac{1}{\sqrt{n}}\sum_{j=[n(k-1)\alpha]}^{[n(k-1)\alpha]+\ell}Y_j\right|\geq\epsilon/2)\leq\frac{1}{n^2\epsilon^4}\mathbb{E}\left(\left(\sum_{j=[n(k-1)\alpha]}^{[n(k-1)\alpha]+[\alpha(n+1)]}Y_j\right)^4\right).$$

On the other hand, one has

$$\mathbb{E}\left(\left(\sum_{j=[n(k-1)\alpha]}^{[n(k-1)\alpha]+[\alpha(n+1)]}Y_j\right)^4\right)\leq C\alpha^2(n+1)^2.$$

Putting together the preceding inequalities we get

$$\mathbb{P}(\sup_{|x-y|<\alpha}|\xi_n(x)-\xi_n(y)|>\epsilon)\leq C\alpha^{-1}\frac{1}{n^2\epsilon^4}\alpha^2(n+1)^2\leq C\alpha\epsilon^{-4}$$

which tends to 0 as α approaches to 0. □

The CLT and the invariance principle are the only examples of limit theorem for which we will give complete proofs. Nevertheless similar computations can provide other properties. Let us mention a couple of examples.

1.3.1.2 The CLT for Vector Valued Functions

Consider the case where φ has values in $\mathbb{R}^2$ (the d-dimensional case is treated in the same manner). We say that $\varphi=(\varphi_1,\varphi_2)$ generates a sequence of differences of a martingale if φ_1 and φ_2 do. Then for all t_1,t_2 in $t_1\varphi_1+t_2\varphi_2$ generates a sequence of differences of a martingale. The CLT for the associated normalized sequence give:

$$\mathbb{E}(\exp(\frac{1}{\sqrt{n}}i\sum_{j=0}^{n-1}(t_1\varphi_1+t_2\varphi_2)\circ T^j))\to_{n\to\infty}\exp(-\frac{\sigma^2(t_1,t_2)}{2}),$$

where

$$\sigma^2(t_1,t_2)=\mathbb{E}((t_1\varphi_1+t_2\varphi_2)^2)=t_1^2\mathbb{E}(\varphi_1^2)+2t_1t_2\mathbb{E}(\varphi_1\varphi_2)+t_2^2\mathbb{E}(\varphi_2^2).$$

This exactly means that we have CLT for vector valued functions: the $\mathbb{R}^2$-valued process $\frac{1}{\sqrt{n}} \sum_{j=0}^{n-1} \varphi \circ T_j$ tends to the gaussian distribution on $\mathbb{R}^2$ with covariance matrix

$$\begin{pmatrix} \mathbb{E}(\varphi_1^2) & \mathbb{E}(\varphi_1\varphi_2) \\ \mathbb{E}(\varphi_1\varphi_2) & \mathbb{E}(\varphi_2^2) \end{pmatrix}.$$

This limit distribution is degenerated if its covariance matrix has determinant 0. Cauchy-Schwarz inequality shows that this is the case only when φ_1 and φ_2 are proportional that is when φ takes its values in a line of $\mathbb{R}^2$.

If we apply the Gordin method to a regular function with values in $\mathbb{R}^2$ then we obtain that the normalized ergodic sums tend to a non degenerated gaussian vector if φ is not cohomologous to a function taking its values in a line of $\mathbb{R}^2$.

We also have a multidimensional version of the Donsker invariance principle that asserts the convergence in distribution of the interpolated lines defined by the ergodic sums of $\mathbb{R}^d$-valued regular functions to a Brownian motion in $\mathbb{R}^d$.

1.3.1.3 The CLT Along Subsequences

We consider ergodic sums at random times with the hypothesis that these times define a sequence of positive density in $\mathbb{N}$.

Theorem 1.8 ("CLT along subsequences"). *Let (Y_k) be a sequence of differences of a martingale defined on a probability space $(\Omega, \mathcal{B}, \mathbb{P})$ and $n : \Omega \to \mathbb{N}$ such that there exists a $\in]0, 1[$ such that, for $\mathbb{P}$-almost ω,*

$$\frac{Card\{n(\omega)/n(\omega) \le k\}}{k} \to_{k \to \infty} a.$$

Then

$$\frac{1}{\sqrt{n}} \sum_{j=0}^{n(\cdot)-1} Y_j(\cdot) \to_{\mathscr{L}} \mathscr{N}(0, a\mathbb{E}(Y_0^2)).$$

1.3.2 Example 3: The Geodesic Flow on a Compact Surface with Curvature −1

The geodesic flow on a compact surface of constant negative curvature -1 is algebraically defined as the action on the quotient of $G = PSL(2, \mathbb{R})$ by a cocompact lattice Γ of the one parameter group

$$g_t : G/\Gamma \longrightarrow G/\Gamma$$

$$y \longmapsto \begin{pmatrix} e^{t/2} & 0 \\ 0 & e^{-t/2} \end{pmatrix} y.$$

The horocyclic flows are defined by the two one parameter groups $(h_t^u)_{t \in \mathbb{R}}$ and $(h_t^s)_{t \in \mathbb{R}}$:

$$h_t^u = \begin{pmatrix} 1 & t \\ 0 & 1 \end{pmatrix},$$

$$h_t^s = \begin{pmatrix} 1 & 0 \\ t & 1 \end{pmatrix}.$$

The orbits of the groups $(h_t^u)_{t \in \mathbb{R}}$ and $(h_t^s)_{t \in \mathbb{R}}$ give respectively the unstable and stable leaves of the geodesic flow:

$$g_t h_a^u g_{-t} = h_{e^t a}^u, \ g_t h_a^s g_{-t} = h_{e^{-t} a}^s, \ \forall t, a \in \mathbb{R}.$$

We are interested in the convergence in distribution of integrals of the form

$$\int_0^T f(g_t \cdot) dt.$$

A simple computation shows that under the condition

$$\int_0^\infty |\langle f(g_t \cdot), f \rangle| dt < \infty$$

the limit

$$\sigma^2(f) = \lim_T \frac{1}{T} \| \int_0^T f(g_t \cdot) dt \|_{L^2(m)}^2$$

exists and equals

$$\sigma^2(f) = \int_{-\infty}^\infty \langle f(g_t \cdot), f \rangle dt.$$

Thus in these cases the normalization leading to a convergence in distribution should be to divide $\int_0^T f(g_t \cdot) dt$ by $\sqrt{T}$.

To construct a filtration as explained before, it is convenient to be in a discrete setting. Let T denotes the translation by g_1 to the left:

$$T : G/\Gamma \longrightarrow X = G/\Gamma$$

$$y \longmapsto \begin{pmatrix} e^{1/2} & 0 \\ 0 & e^{-1/2} \end{pmatrix} y.$$

We get the discrete dynamical system (X, T, μ) where $X = G/\Gamma$ and μ is the probability on X deduced from the Haar measure on G.

Let us explain how to construct a convenient filtration.

We will use the following property of equidistribution of unstable leaves for the geodesic flow, due to Burger [10]. It is a strong result that is not necessary for the first steps of the constructions but that will be crucial to establish the convergence of one of the first series when applying Theorem 1.4.

Theorem 1.9. *There exist $\alpha > 0$ and $C > 0$ such that, for every C^∞-function f on $X = G/\Gamma$, for every x and $t \geq 1$, one has:*

$$\left| \frac{1}{T} \int_0^T f(h_t^u x) dt - \int_{G/\Gamma} f(y) d\mu(y) \right| \leq CT^{-\alpha} \|f\|_{H_2^3},$$

where $\|f\|_{H_2^3}$ refers to some Sobolev norm.

This enables us to construct an adequate filtration. Let $r > 0$ be a number and z_0 a point in X such that the "cube"

$$\mathscr{P}_0 = \{h_v^u g_t h_w^s z_0 \, / \, (v, t, w) \in [-r, r]^3\}$$

do not coincide with X (it is the case if r is small enough). The set $\mathscr{P}_0$ defines a partition of X: $\mathscr{P}_0$ and its complementary set. They both have a non-empty interior. Because of the result of Burger, every orbit of $(h_t^u)_{t \in \mathbb{R}}$ do cross $\mathscr{P}_0$ and $^c\mathscr{P}_0$. It means in particular that every orbit of $(h_t^u)_{t \in \mathbb{R}}$ hit the face

$$\mathscr{F}_0 = \{g_t h_w^s z_0 \, / \, (t, w) \in [-r, r]^2\}.$$

For every x in $X \setminus \mathscr{F}_0$, there exist two numbers $l_x < 0$ and $u_x > 0$ such that:

$$h_{l_x}^u x \in \mathscr{F}_0, \ h_{u_x}^u x \in \mathscr{F}_0, \ \forall t \in (l_x, u_x), h_t^u x \notin \mathscr{F}_0.$$

In this way we define the partition $\mathscr{Q}$ the atoms of which are the sets

$$\mathscr{Q}(x) = \{h_t^u x \, / \, t \in (l_x, u_x)\}.$$

This is a partition up to the null set $\mathscr{F}_0$. We then define the partition $\mathscr{Q}_0^\infty$:

$$\mathscr{Q}_0^\infty(x) = \bigcap_{k \geq 0} T^k \mathscr{Q}(T^{-k} x).$$

We have

$$T \mathcal{Q}_0^\infty(x) = \bigcap_{k \geq 0} T^{k+1} \mathcal{Q}(T^{-k}x) = \bigcap_{k \geq 1} T^k \mathcal{Q}(T^{-k}Tx)$$

so that

$$\mathcal{Q}_0^\infty(Tx) = \mathcal{Q}(Tx) \bigcap T \mathcal{Q}_0^\infty(x).$$

This means that the atoms of $\mathcal{Q}_0^\infty$ have the following properties:

- They are finite pieces of orbits of $(h_t^u)_{t \in \mathbb{R}}$.
- The image by T of an atom of $\mathcal{Q}_0^\infty$ is a union of atoms.

Let $\mathcal{A}_0$ be the σ-algebra of Borel sets that are saturated for the equivalence relation defined by $\mathcal{Q}_0^\infty$. The second of the preceding properties shows that $T\mathcal{A}_0$ is a sub-σ-algebra of $\mathcal{A}_0$. For every $k \in \mathbb{Z}$, let $\mathcal{A}_k$ denotes the σ-algebra $T^{-k}\mathcal{A}_0$: we obtain an increasing filtration of σ-algebras the atoms of which are finite pieces of orbits of $(h_t^u)_{t \in \mathbb{R}}$.

We now want a control of the lengths of the atoms of $\mathcal{A}_0$. Let $\epsilon > 0$: when is the piece $\{h_t^u x \ / \ t \in [-\epsilon, \epsilon]\}$ included in $\mathcal{Q}_0^\infty(x)$? When $\{h_t^u x \ / \ t \in [-\epsilon, \epsilon]\}$ is included in $T^k \mathcal{Q}(T^{-k}x)$ for every $k \geq 0$, that is when

$$T^{-k}\{h_t^u x \ / \ t \in [-\epsilon, \epsilon]\} = \{h_{e^{-k}t}^u T^{-k}x \ / \ t \in [-\epsilon, \epsilon]\}$$

does not hit $\mathcal{F}_0$. In other words $\{h_t^u x \ / \ t \in [-\epsilon, \epsilon]\}$ is included in $\mathcal{Q}_0^\infty(x)$, if for every $k \geq 0$, $T^{-k}x$ does not belongs to the set $\{h_t^u y \ / \ y \in \mathcal{F}_0, t \in [-e^{-k}\epsilon, e^{-k}\epsilon]\}$. These sets have a measure bounded by $Ce^{-k}\epsilon$. As T preserves the measure m, we get the proposition:

Proposition 1.2. *There exists $C > 0$ such that the set of points x for which $\{h_t^u x \ / \ t \in [-\epsilon, \epsilon]\}$ is included in $\mathcal{Q}_0^\infty(x)$ has measure larger than $1 - C\epsilon$.*

Proposition 1.3. *The conditional expectation of a function f with respect to $\mathcal{A}_0$ is almost surely given by*

$$\mathbb{E}(f|\mathcal{A}_0)(x) = \frac{1}{u_x^\infty - l_x^\infty} \int_{l_x^\infty}^{u_x^\infty} f(h_t^u x)dt,$$

where $\mathcal{Q}_0^\infty(x) = \{h_t^u x \ / \ t \in (l_x^\infty, u_x^\infty)\}$.

We are now ready to prove the convergence of the series

$$\sum_{n<0} ||E(f|\mathcal{A}_n)||_2 \quad \text{and} \quad \sum_{n>0} ||f - E(f|\mathcal{A}_n)||_2$$

for a C^∞ function. The second convergence is an obvious consequence of the mean value theorem: the size of the atoms of $\mathscr{A}_n$ are uniformly exponentially small for n tending toward infinity. To get the first one we proceed as follows: let $\beta < e$, the set V_n of points x for which $\{h_t^u x / t \in [-\beta^{-n}, \beta^{-n}]\}$ is included in $\mathscr{Q}_0^\infty(x)$ has measure larger than $1 - C\beta^{-n}$. If x belongs to $T^n V_n$, then the atom $\mathscr{A}_{-n}(x) = T^n \mathscr{Q}_0^\infty(T^{-n}x)$ is a piece of orbit of $(h_t^u)_{t \in \mathbb{R}}$ of length larger than $2e^n \beta^{-n}$; the result of Burger gives

$$|\mathbb{E}(f|\mathscr{A}_{-n})(x)| \leq C(2e^n\beta^{-n})^{-\alpha}\|f\|_{H_2^3}, \text{ for } x \in T^n V_n.$$

Thus we have

$$\|E(f|\mathscr{A}_{-n})\|_2 \leq C(2e^n\beta^{-n})^{-\alpha}\|f\|_{H_2^3} + \mu(^cT^nV_n)^{1/2}\|f\|_\infty$$
$$\leq C\left((e^n\beta^{-n})^{-\alpha} + \beta^{-n/2}\right),$$

the convergence of the first series follows.

The result still holds for Hölder continuous functions. To establish it, one has to regularize it using convolution.

This technique can be adapted to the finite volume non compact case. Some changes are needed: one knows that some little periodic horocycles around a cusp won't hit a given two dimensional compact rectangle placed in G/Γ.

1.3.3   Example 4: The Ergodic Automorphisms of the Torus

Let M be a square integer matrix without an eigenvalue which is a root of unity and of determinant ± 1. This matrix necessarily has eigenvalues inside and outside the unit circle. This matrix defines a transformation of the torus

$$T : \mathbb{T}^d \to \mathbb{T}^d \quad x \mapsto Mx \bmod \mathbb{Z}^d.$$

This transformation preserves the Lebesgue measure and the associated dynamical system $(\mathbb{T}^d, T, \mu)$ is ergodic.

Let F_u (resp. F_s, resp. F_e) be the M-stable vector space associated to the eigenvalues of modulus larger than (resp. smaller than, resp. equal to) 1. Let $v_1, \ldots, v_d$ be a basis of $\mathbb{R}^d$ in which M is represented by a real Jordan matrix. Let r denote the dimension of F_u and suppose that $v_1, \ldots, v_r$ is a basis of F_u. Consider a partition $\mathscr{Q}$ of the torus whose elements are of sufficiently small diameter and of the form $\sum I_i v_i$, where the I_i are intervals. As in the preceding section we then define the partition $\mathscr{Q}_0^\infty$:

$$\mathscr{Q}_0^\infty(x) = \bigcap_{k \geq 0} T^k \mathscr{Q}(T^{-k}x),$$

so that

$$\mathcal{Q}_0^\infty(Tx) = \mathcal{Q}(Tx) \bigcap T\mathcal{Q}_0^\infty(x).$$

Let $\mathscr{A}_0$ be the σ-algebra of borelian sets that are saturated for the equivalence relation defined by $\mathcal{Q}_0^\infty$. The image $T\mathscr{A}_0$ is a sub-σ-algebra of $\mathscr{A}_0$. For every $k \in \mathbb{Z}$, let $\mathscr{A}_k$ denotes the σ-algebra $T^{-k}\mathscr{A}_0$: we obtain an increasing filtration of σ-algebras the atoms of which are (uniformly) bounded convex pieces (F_u).

The partition $\mathcal{Q}_0^\infty$ is measurable in the sense of Rokhlin: for almost every atom $\mathscr{A}_n(x)$ of $\mathscr{A}_n$, we can define a conditional probability $\mu_{\mathscr{A}_n(x)}$ on $\mathscr{A}_n(x)$ such that, for m-almost every x, one has

$$E\left(f \mid \mathscr{A}_n\right)(x) = \int_{\mathscr{A}_n(x)} f(t)\, d\mu_{\mathscr{A}_n(x)}(t).$$

For almost every x, the atom $\mathscr{A}_n(x)$ is a set included in $x + F_u$ with non empty interior and the conditional probability is the restriction to the atom of m_u be the Lebesgue measure on F_u. So almost surely we have (remark that $\mathscr{A}_n = T^{-n}\mathscr{A}_0$)

$$E\left(f \mid \mathscr{A}_n\right)(x) = \int_{\mathscr{A}_n(x)} f(t)\, d\mu_{\mathscr{A}_n(x)}(t)$$

$$= \frac{1}{m_u(\mathscr{A}_n(x))} \int_{\mathscr{A}_n(x)} f(t)\, dm_u(t)$$

$$= \frac{1}{m_u(T^{-n}\mathscr{A}_0(T^n x))} \int_{T^{-n}\mathscr{A}_0(T^n x)} f(t)\, dm_u(t).$$

As in the case of the cat map we need a diophantine estimation to get the good repartition of the unstable leaves in the torus. It can be expressed here by saying that the integer points are not very near to the stable eigenspace of the matrix defining our automorphism. A result found in [31] gives a sufficient information: there exist constants $K > 0$ and $q > 0$ such that, for every $k \in \mathbb{Z}^d \setminus \{0\}$, one has

$$\|\pi_u k\| \geq K\|k\|^{-(d-r)},$$

where π_u is the projection on F_u along $F_e + F_s$.

We also take advantage of the convexity of the atoms of $\mathscr{A}_n$. Let C be a compact convex set in F_u. Let $a(C)$ denote the area of the boundary of C, that is the $(r-1)$-dimensional measure of this boundary computed in basis $v_1, \ldots, v_r$. Let C be a convex compact set included in $x + F_u$ with nonempty interior (in F_u) and γ be a positive real number. We have the following inequality

$$m_u\{y \in C \ / \ d(y, \partial C) \leq \gamma\} \leq \gamma a(C).$$

Using the preceding remarks we can show that there exist $K > 0$ and $\lambda \in]0, 1[$ such that, for every x, every convex compact C included in $x + F_u$ with nonempty interior (in F_u) and every $k \in \mathbb{Z}^d \setminus \{0\}$, we have

$$\frac{1}{m_u(T^{-n}C)} \left| \int_{T^{-n}C} \exp(2i\pi\langle k, t\rangle) \, dm_u(t) \right| < K \frac{a(C)}{m_u(C)} \|k\|^{d-r} \lambda^n.$$

Let f a C^∞ function. Using the Fourier series (and the fact that the Fourier coefficients of f rapidly decrease) it is the easy to prove that the two series

$$\sum_{n>0} \|f - E(f|\mathscr{A}_n)\|_2$$

and

$$\sum_{n<0} \|E(f|\mathscr{A}_n)\|_2$$

converge.[3]

As we have seen on the examples considered till now, it is easy to construct a filtration $\mathscr{A}_n$ the atoms of which are pieces of unstable leaves. When it is done, for Lipschitz continuous functions f, the convergence of the series $\sum_{n>0} \|f - E(f|\mathscr{A}_n)\|_2$ is obvious. The only serious problem is to get convergence of the other series $\sum_{n<0} \|E(f|\mathscr{A}_n)\|_2$. Two difficulties do appear: to control the shape of the pieces, to establish the good repartition of large pieces of stable leaves with a boundary that is not to complicated. In the two previous examples representation theory or Fourier series provided the second information. The first one was obtained through convexity or because of the dimension 1 of the stable leaves. We will now show how the method works in other algebraic situations.

1.4 Martingales in Hyperbolic Geometry

1.4.1 Example 5: The Geodesic Flow in Dimension d, Constant Curvature (Compact Case)

We consider the geodesic flow on a d-dimensional compact hyperbolic manifold of constant negative curvature. This flow is algebraically defined as follows.[4] Consider the group $SO(1, d)$ of the matrices that leave invariant the quadratic form:

---

[3]For more details see [33].

[4]We adopt some of the notations of [20].

$$\langle x, x \rangle = x_0^2 - \sum_{k=1}^{d} x_k^2$$

defined on $\mathbb{R}^{d+1}$. This group has two connected components. The component of the identity is a subgroup of index two denoted $PSO(1, d)$. The unit tangent bundle of a compact manifold of constant negative curvature M can be identified with a quotient $SO(d-1)\backslash PSO(1, d)/\Gamma$ where $SO(d-1)$ is identified with the matrices

$$\begin{pmatrix} 1 & 0 & 0 \\ 0 & 1 & 0 \\ 0 & 0 & R \end{pmatrix}$$

with $R \in SO(d-1)$ and Γ is a cocompact discrete subgroup of $PSO(1, d)$. The geodesic flow is defined by the one parameter group $(g_t)_{t \in \mathbb{R}}$ where the g_t are the matrices

$$g_t = \begin{pmatrix} \cosh(t) & \sinh(t) & 0 & 0 & \dots & 0 & 0 \\ \sinh(t) & \cosh(t) & 0 & 0 & \dots & 0 & 0 \\ 0 & 0 & 1 & 0 & \dots & 0 & 0 \\ 0 & 0 & 0 & 1 & & 0 & 0 \\ \vdots & & \vdots & \vdots & \vdots & 1 & 0 \\ 0 & 0 & 0 & 0 & \dots & 0 & 1 \end{pmatrix}.$$

This action commutes with the one of matrices

$$\begin{pmatrix} 1 & 0 & 0 \\ 0 & 1 & 0 \\ 0 & 0 & R \end{pmatrix}$$

so that it is well defined on $SO(d-1)\backslash PSO(1, d)/\Gamma$. The action of $(g_t)_{t \in \mathbb{R}}$ on $PSO(1, d)/\Gamma$ is the action on the so called frame bundle. In the sequel g_t denotes the matrix or the transformation it defines on $X = PSO(1, d)/\Gamma$ by

$$g_t : PSO(1, d)/\Gamma \longrightarrow PSO(1, d)/\Gamma \ : \ x \longmapsto g_t x.$$

As in Sect. 1.3.2 we work in the discrete time dynamical system $(G/\Gamma, T = g_1, \mu)$ and we treat only the case of C^∞ functions. The **variance** $\sigma^2(\varphi)$ is the limit (when it exists and is finite):

$$\sigma^2(\varphi) = \lim_{n \to +\infty} \frac{1}{n} \|\varphi + T\varphi + \dots + T^{n-1}\varphi\|_2^2.$$

Let $S_n\varphi$ be the $\varphi + T\varphi + \ldots + T^{n-1}\varphi$.

Like in dimension 2 we have horogroups that are expanded or contracted by g_t: for $v, w \in \mathbb{R}^{d-1}$, $v = (v_2, \ldots, v_d)$, $w = (w_2, \ldots, w_d)$

$$\theta_v^+ = \begin{pmatrix} 1 + \frac{|v|^2}{2} & -\frac{|v|^2}{2} & v_2 \, v_3 \, \ldots \, v_{d-1} \, v_d \\ \frac{|v|^2}{2} & 1 - \frac{|v|^2}{2} & v_2 \, v_3 \, \ldots \, v_{d-1} \, v_d \\ v_2 & -v_2 & 1 \, 0 \, \ldots \, 0 \, 0 \\ v_3 & -v_3 & 0 \, 1 \qquad 0 \, 0 \\ \vdots & \vdots & \vdots \, 0 \, \ddots \qquad 0 \\ \vdots & \vdots & \vdots \, \vdots \qquad 1 \, 0 \\ v_d & -v_d & 0 \, 0 \, \ldots \, 0 \, 1 \end{pmatrix}$$

and $\theta_w^- = {}^t\theta_w^+$ the transpose of θ_w^+. A simple computation shows that we have the relations

$$\theta_v^+ \theta_{v'}^+ = \theta_{v+v'}^+, \quad \theta_w^- \theta_{w'}^- = \theta_{w+w'}^-,$$

$$g_t \theta_v^+ g_{-t} = \theta_{e^t v}^+, \quad g_t \theta_w^- g_{-t} = \theta_{e^{-t} w}^-.$$

Let us denote K the compact subgroup identified with $SO(d-1)$ and θ^+ (resp. θ^-) the groups isomorphic to $\mathbb{R}^{d-1}$ of matrices θ_v^+ (resp. θ_w^+). The stable and unstable leaves for the action on the frame bundle are thus very simple: they are described by the orbits of the group $\mathbb{R}^{d-1}$ (through θ^+ and θ^-) on $PSO(1, d)/\Gamma$. On the level of the tangent bundle $SO(d-1)\backslash PSO(1, d)/\Gamma$, the description of these leaves are more complicated. Remark that at this level the action of $(g_t)_{t \in \mathbb{R}}$ defines an Anosov flow: it gives the possibility to apply different methods and obtain strong results among which the CLT. We will work at the level of G/Γ. The method of martingales gives the CLT for the action on the frame bundle.

The Liouville measure corresponds to the measure on $X = PSO(1, d)/\Gamma$ deduced from the Haar measure on $PSO(1, d)$. Let μ be this measure normalized so that $\mu(X) = 1$. The four groups k ($\in SO(d-1)$), θ_v^+, g_t, θ_w^- define local systems of coordinates:

$$(v, t, k, w) \mapsto \theta_v^+ g_t k \theta_w^- x, \ (v, t, w) \in [-2r_0, 2r_0]^{2d-1}, k \in B_{SO(d-1)}(2r_0)$$

$$(v, t, k, w) \mapsto \theta_w^- g_t k \theta_v^+ x, \ (v, t, w) \in [-2r_0, 2r_0]^{2d-1}, k \in B_{SO(d-1)}(2r_0)$$

for r_0 small enough. The expression of the measure μ in these systems of coordinates is given by

$$\int f(gx) d\mu(g) = \int_{[-r_0, r_0]^{2d-1} \times B_{SO(d-1)}(\delta)} f(\theta_v^+ g_t k \theta_w^- x) e^{(d-1)t} \, dv dt dw dk.$$

We cover X with a finite family $\mathcal{R}$ of boxes of the form

$$(v, t, k, w) \mapsto \theta_v^+ g_t k \theta_w^- x, \ (v, t, w) \in [-r_0, r_0]^{2d-1}, k \in B_{SO(d-1)}(r_0).$$

It induces a finite partition $\mathcal{P}$, which we cut into slices parametrized by θ^+ to get a (infinite, non denumerable) partition $\mathcal{Q}$:

$$\mathcal{P} = \{\bigcap_{R \in \mathcal{R}} A_R \ / \ A_R = R \text{ or } {}^c R\},$$

$$\mathcal{Q}(x) = \mathcal{P}(x) \bigcap \theta_{[-2r_0, 2r_0]^{d-1}}^+ x.$$

If a point y lies in two boxes

$$y = \theta_v^+ g_t k \theta_w^- x = \theta_{v'}^+ g_{t'} k' \theta_{w'}^- \overline{x'}$$

then $\theta_u^+ y$ is still in the two boxes if $u + v$ and $u + v'$ belongs to $[-r_0, r_0]^{d-1}$. From this we deduce that the atoms of $\mathcal{Q}$ have the following form:

$$\mathcal{Q}(y) = \{\theta_u^+ y \ / \ u \in \prod_{k=2}^{d} [\alpha_l(k, y), \alpha_u(k, y)]\}.$$

As before, let us then define

$$\mathcal{Q}_0^\infty(x) = \mathcal{Q}(x) \cap T \mathcal{Q}(T^{-1}x) \cap T^2 \mathcal{Q}(T^{-2}x) \cap \ldots$$

This is the coarser partition among those which are finer than $\mathcal{Q}$ and that are included in their image by T^{-1}.

For every $n \in \mathbb{Z}$ let us call $\mathcal{A}_n$ the σ-algebra the atoms of which are the sets $T^{-n} \mathcal{Q}_0^\infty$. The sequence $(\mathcal{A}_n)$ is an increasing filtration. The atom of $\mathcal{A}_n$ containing x is

$$\mathcal{A}_n(x) = T^{-n} \mathcal{Q}(T^n x) \cap T^{-(n-1)} \mathcal{Q}(T^{n-1}x) \cap T^{-(n-2)} \mathcal{Q}(T^{n-2}x) \cap \ldots$$

$$= T^{-n} \mathcal{A}_0(T^n x).$$

The equality $\mathcal{A}_1(x) = T^{-1} \mathcal{Q}(Tx) \cap \mathcal{A}_0(x)$ shows that the atoms of $\mathcal{A}_0$ are unions of atoms of $\mathcal{A}_1$.

As in the preceding examples, consider the set $W_n^{\delta,\beta}$ ($\delta, \beta \in]1, \infty[$)

$$W_n^{\delta,\beta} = \{x \in G/\Gamma \ / \ \forall k \geq 0 \ \Delta^+(\beta^{-n}\delta^{-k})T^{-k}x \subset \mathcal{Q}(T^{-k}x)\}.$$

We have $T^1 \Delta^+(r) T^{-1} = \Delta^+(er)$.

If $1 < \delta < e$ then, for x in $W_n^{\delta,\beta}$, for every $k \geq 0$, we have

$$\Delta^+(\beta^{-n})x \subset \Delta^+(\beta^{-n}e^k\delta^{-k})x \subset T^k \Delta^+(\beta^{-n}\delta^{-k})T^{-k}x,$$

thus

$$\Delta^+(\beta^{-n})x \subset \mathscr{Q}_0^\infty(x).$$

On the other hand, as the boundaries of $\mathscr{Q}$ are regular, there exists $C > 0$ such that, for every $\epsilon > 0$, we have

$$\mu\{x \in G/\Gamma \ : \ \Delta^+(\epsilon)x \cap \partial\mathscr{Q}(x) \neq \emptyset\} \leq C\epsilon.$$

This allows us to get the following lemma:

Lemma 1.2. *There exists $C > 0$ and $q > 0$ such that*

$$\mu(^cW_n^{\delta,\beta}) \leq C\beta^{-n}.$$

Proof. This is a consequence of the computations (we use the invariance of the measure at the second line)

$$\mu(^cW_n^{\delta,\beta}) \leq \sum_{k=0}^\infty \mu\{x/\Delta^+(\beta^{-n}\delta^{-k})T^{-k}x \cap \partial\mathscr{Q}(T^{-k}x) \neq \emptyset\}$$

$$\leq \sum_{k=0}^\infty \mu\{x/\Delta^+(\beta^{-n}\delta^{-k})x \cap \partial\mathscr{Q}(x) \neq \emptyset\}$$

$$\leq \sum_{k=0}^\infty C\beta^{-n}\delta^{-k}. \qquad \square$$

The atoms of $\mathscr{Q}_0^\infty$ contain with large probability a piece of unstable leaf of size larger than β^{-n}.

To get the CLT we have to prove that large unstable pieces of unstable leaves are well distributed in X. To say it otherwise we need a quantitative information on the equidistribution of the horospheres in X. We will get it through the mixing properties of g_t. In our algebraic setting these mixing properties are the decreasing properties of matrix coefficients of unitary representations of $PSO(1, d)$. Let us describe briefly a classical way to deduce a mixing property for some differentiable functions. It is similar to the idea of integrations by parts that give information on the decay of the Fourier coefficients of a several times differentiable periodic function. This idea still works in a more general setting, but one has to replace the imaginary exponential functions by what is called K_m-finite functions (where K_m is a maximal compact subgroup of G), and the differentiation by the application of the Casimir operator.

Consider π the restriction to $L^2_0(G/\Gamma)$ (the centered functions) of the natural representation of G on $L^2(G/\Gamma)$. It is proved in [3] that π does not weakly contain the identity. Using [40] we deduce that π is strongly L^p for some L^p. Then [27] gives the following theorem.

Theorem 1.10. *There exists $\zeta > 1$ such that for every K_m-finite vectors φ, ψ in $L^2_0(G/\Gamma)$ and every $t \in \mathbb{R}$ one has*

$$|\langle g_t \varphi, \psi \rangle| \leq C \dim_{K_m}(\varphi)^{\frac{1}{2}} \dim_{K_m}(\psi)^{\frac{1}{2}} \|\varphi\| \|\psi\| \zeta^{-t}.$$

It states the exponential decay of correlations for some particular functions: the K_m-finites ones, that is the functions φ for which the vector space generated by the action of K_m (by translation to the left: $\varphi(k^{-1}\cdot)$) has finite dimension (denoted $\dim_{K_m}(\varphi)$). We explain how to pass from K_m-finite to C^∞ functions (see [30]).

The action of K_m on X defines a unitary representation U of K_m on $L^2(\mu)$ by $k \longmapsto \varphi(k^{-1}x)$ which can be decomposed as a sum of irreducible representations. Let $\hat{K}_m$ be the set of the equivalence classes of irreducible representations of K_m and δ be an element of $\hat{K}_m$.

Let us fix a base R of the root system of $\mathscr{K}_m$ the Lie algebra of K_m. Let us call W the associated Weyl chamber. To each irreducible representation of K_m is uniquely associated a linear form belonging to a lattice in W: the dominant weight of the representation. Let δ be an element of $\hat{K}_m$ and let γ_δ be the corresponding dominant weight.

The Weyl formula gives the dimension d_δ of the irreducible representation associated to δ as a function of γ:

$$d_\delta = \prod_{\alpha \in R_+} \frac{\langle \alpha, \gamma_\delta + \rho \rangle}{\langle \alpha, \rho \rangle},$$

where R_+ is the set of positive roots and ρ the half sum of the positive roots.

For every $\delta \in \hat{K}_m$, let ξ_δ be the character of δ (the function on K_m defined as the trace of the action of the elements of K_m on the irreducible representation of K_m associated to δ), $\chi_\delta = d_\delta \xi_\delta$, and

$$P_\delta = U(\overline{\chi_\delta}) = d_\delta \int_{K_m} \overline{\xi_\delta(k)}\, U(k)\, dk. \tag{1.1}$$

Remark that $\|P_\delta \varphi\|_\infty \leq d_\delta^2 \|\varphi\|_\infty$. The operator P_δ is the projection of $L^2(\mu)$ on the isotypic part $\mathscr{F}_\delta := P_\delta(L^2(\mu))$. We have the decomposition

$$L^2(\mu) = \bigoplus_{\delta \in \hat{K}_m} \mathscr{F}_\delta.$$

For a given vector v in $L^2(\mu)$ let $v_\delta := P_\delta v$. An element v of $\mathscr{F}_\delta$ is K_m-finite:

$$dim \ \mathrm{Vect} K_m v \leq d_\delta^2. \tag{1.2}$$

One says that v is C^∞ if the map $k \mapsto U(k)v$ is C^∞. One defines the derived representation of U on the space of C^∞ elements; it is a representation of the Lie algebra $\mathscr{K}_m$ of K_m and that one can be extended to a representation of the universal enveloping algebra of $\mathscr{K}_m$. We use the same later U to denote these three representations.

Let $X_1, \ldots, X_n$ ($n = d(d-1)/2$) be an orthonormal basis for an invariant scalar product on $\mathscr{K}_m$. The operator $\Omega = 1 - \sum_{i=1}^n X_i^2$ belongs to the center of the universal enveloping algebra of $\mathscr{K}_m$. So, by Schur's lemma, if μ_δ is a representation of the type δ, there exists c_δ such that $\mu_\delta(\Omega) = c_\delta \mu_\delta(1)$.

The operators Ω being hermitian, c_δ is positive. One can show (cf. [6]) that there exists a scalar product Q such that $c_\delta = Q(\gamma_\delta + \rho) - Q(\rho)$.

If v is C^∞, one has

$$P_\delta U(\Omega)v = c_\delta P_\delta v = c_\delta v_\delta,$$

thus, for every non negative integer m, for every δ in $\hat{K}_m$, one has

$$v_\delta = c_\delta^{-m} (U(\Omega^m)v)_\delta.$$

From this equality and the definition (1.1) of P_δ, one deduces that

$$\|v_\delta\|_\infty \leq \frac{d_\delta^2}{c_\delta^m} \|U(\Omega^m)v\|_\infty.$$

One can chose m so large that the series

$$\sum_{\delta \in \hat{K}} \frac{d_\delta^2}{c_\delta^m}$$

converges. For such a choice of m we have the following proposition.

Proposition 1.4. *Let φ and ψ be two centered C^∞ functions on G/Γ. There exist two numbers $\zeta > 1$ and $C > 0$ and a differential operator Ω such that, for every n, one has:*

$$|\langle \varphi, \psi \circ T^n \rangle| \leq C \|\Omega^m \varphi\|_2 \|\Omega^m \psi\|_2 \zeta^{-|n|}. \tag{1.3}$$

Let us come back to the CLT. Here we consider the time one map $T = g_1$ of the flow g_t. The theorem for continuous time follows.

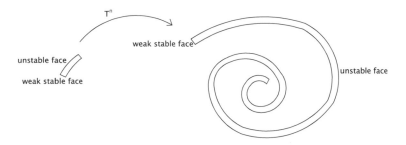

Fig. 1.2 Coiling of the image of a "cube" in X

As already said our proof of the CLT will rely on a quantitative information of the repartition of the unstable leaves of T in G/Γ. We obtain it through the exponential decay of correlations for T. This link between decorrelation and equidistribution of unstable leaves has been used by different authors (cf. for example [32]).

The link between decorrelation and equirepartition is easy to understand. We will use the following notations:

$$\Delta^-(\epsilon) = \theta^-_{[-\epsilon,\epsilon]^{d-1}} = \{\theta^-_w \ / \ w \in [-\epsilon,\epsilon]^{d-1}\}$$

$$\Delta^+(\epsilon) = \theta^+_{[-\epsilon,\epsilon]^{d-1}} = \{\theta^+_w \ / \ w \in [-\epsilon,\epsilon]^{d-1}\}$$

and $B_K(\epsilon)$ be the set $K \cap B(Id, \epsilon)$. Fix an $\epsilon > 0$ and consider a set $F \subset \theta^+$ containing the set $\Delta^+(\epsilon)$. Let U be the set $\theta^-_{[\epsilon,\epsilon]^{d-1}} g([\epsilon,\epsilon]) B_K(Id,\epsilon) F$. As T contracts θ^- by e and commutes with K and g, we have

$$T^n U T^{-n} = \theta^-_{[e^{-n}\epsilon, e^{-n}\epsilon]^{d-1}} g([\epsilon,\epsilon]) B_K(Id,\epsilon) T^n F T^{-n}.$$

Take $\epsilon > 0$ small enough so that, the map

$$(v,t,k,w) \longmapsto \theta^-_v g_t k \theta^+_w x$$

is a diffeomorphism from U on its image in $X = PSO(1,d)/\Gamma$. Under the action of T^n this set dilates in the θ^+ direction, coils in X (see Fig. 1.2) and, as we will see, spreads itself uniformly in X.

We can write:

$$\langle \ T^{-n} 1_U, \varphi \rangle$$

$$= \int_{T^n U} \varphi(y) \, d\mu(y)$$

$$= \int_{[-e^{-n}\epsilon, e^{-n}\epsilon]^{d-1} \times [-\epsilon,\epsilon] \times B_K(\epsilon) \times T^n F T^{-n}} \varphi(\theta^-_v g_t k \theta^+ T^n x) \, e^{(d-1)t} \, dv dt dk d\theta^+.$$

For the points involved in this integral, there exists a constant $C > 0$ such that:

$$d(\theta_v^- g_t k \theta^+ T^n x, \theta^+ T^n x) \le C\epsilon.$$

As φ is C^∞, for these points we have:

$$|\varphi(\theta_v^- g_t k \theta^+ T^n x) - \varphi(\theta^+ T^n x)| \le C\epsilon$$

thus,

$$|\langle T^{-n} 1_U, \varphi \rangle - \int_{[-e^{-n}\epsilon, e^{-n}\epsilon]^{d-1} \times [-\epsilon, \epsilon] \times B_K(\epsilon) \times T^n FT^{-n}} \varphi(\theta^+ T^n x) \, e^{(d-1)t} \, dv \, dt \, d\theta^+ |$$

$$\le C\epsilon\mu(U).$$

Dividing by $\mu(U)$ (which is larger than $c\epsilon^D$ for some $c > 0$, $D = \frac{d(d+1)}{2}$ the dimension of $PSO(1, d)$) we obtain:

$$\left| \frac{1}{m_u(T^n FT^{-n})} \int_{T^n FT^{-n}} \varphi(\theta^+ T^n x) \, d\theta^+ \right| \le C \left(\frac{|\langle T^{-n} 1_U, \varphi \rangle|}{\epsilon^D} + \epsilon \right). \qquad (1.4)$$

We just have to bound the quantity $|\langle T^{-n} 1_U, \varphi \rangle|$ which is small because of the exponential decay of correlations. As 1_U is not differentiable, one has to regularize it. For a real number ρ larger than 1, we call ρ-identity a sequence (χ_n) of C^∞ functions defined on $G = PSO(1, d)$, non negative, of integral 1, and such that there exists $C > 0$ for which, for every n:

- the support of χ_n is included in $B(Id, \rho^{-n})$.
- $\|\Omega^m \chi_n\|_\infty \le C\rho^{Cn}$.
- χ_n is Lipschitz-continuous with Lipschitz constant ρ^{Cn}.

Such sequences do exist. For a locally integrable function ψ on $G = PSO(1, d)$, let us consider the convolution:

$$\chi_n * \psi(x) = \int_G \psi(g^{-1}x) \chi_n(g) \, dg = \int_G \psi(g) \chi_n(xg^{-1}) \, dg.$$

We identify the Lie-algebra of G and the set of **right**-invariant vector field on G. Then, for every n, the functions $\chi_n * \psi$ are C^∞ and for every differential operator Ω of the universal enveloping algebra of G, $\Omega(\chi_n * \psi) = (\Omega \chi_n) * \psi$. If φ is an integrable function on G/Γ, it defines a function on G that is Γ invariant. One check that $\chi_n * \varphi$ is right Γ-invariant. Let $\partial U(\beta)$ be the set of points of G/Γ at distance less than β from the boundary of U. Let φ be a C^∞ centered function on G/Γ.

The mixing inequality (1.3) applied to $\chi_n^{(\rho)} * 1_U$ and φ and the properties of the sequence $\chi_n^{(\rho)}$ insure the existence of a constant $C > 0$ such that:

$$\left| \langle T^{-n} \left(\chi_n^{(\rho)} * 1_U \right), \varphi \rangle \right| \leq C \left\| \Omega^m \chi_n^{(\rho)} * 1_U \right\|_2 \| \Omega^m \varphi \|_2 \zeta^{-n} \leq C \| \Omega^m \varphi \|_2 \rho^{Cn} \zeta^{-n}.$$

On the other side we have:

$$\left| \langle 1_U \circ T^{-n}, \varphi \rangle - \langle (\chi_n^{(\rho)} * 1_U) \circ T^{-n}, \varphi \rangle \right| \leq \| \varphi \|_\infty \mu(\partial U(\rho^{-n})).$$

Using these inequalities and (1.4) we get

$$\left| \frac{1}{m_u(T^n F T^{-n})} \int_{T^n F T^{-n}} \varphi(\theta^+ T^n x) \, d\theta^+ \right| \leq C \left(\frac{\mu(\partial U(\rho^{-n})) + \zeta^{-n} \rho^{Cn}}{\epsilon^D} + \epsilon \right). \quad (1.5)$$

We will give convenient values to ϵ and ρ to get the following theorem.

Theorem 1.11. *Let φ a C^∞ centered function on G/Γ. If φ is not a coboundary, then it satisfies the Donsker invariance principle.*

Proof. We want to show the convergence of the two series:

$$\sum_{n>0} \| E(\varphi | \mathscr{A}_n) \|_2 < \infty \quad \text{and} \quad \sum_{n<0} \| \varphi - E(\varphi | \mathscr{A}_n) \|_2 < \infty.$$

The partitions $\mathscr{Q}_n^\infty = T^{-n} \mathscr{Q}_0^\infty$ are measurable in the sense of Rokhlin. For every n, there exists conditional probabilities $(m_P)_{P \in \mathscr{P}_n^\infty}$ which give the values of the conditional expectations with respect to $\mathscr{A}_n$: for every integrable function f, for almost every x,

$$\mathbb{E}(f | \mathscr{A}_n)(x) = \int_{\mathscr{Q}_n^\infty(x)} f(y) \, dm_{\mathscr{Q}_n^\infty(x)}(y).$$

By construction, we have a family $\{F(n, x)\}_{n \in \mathbb{Z}, x \in G/\Gamma}$ of subset of θ^+ such that

$$\mathscr{A}_n(x) = F(n, x)x = T^{-n} \mathscr{A}_0(T^n x) = T^{-n} F(0, T^n x) T^n x,$$

and:

$$\mathbb{E}(\varphi | \mathscr{A}_n)(x) = \frac{1}{m_+(T^{-n} F(0, T^n x) T^n)} \int_{T^{-n} F(0, T^n x) T^n} \varphi(\theta^+ x) \, d\theta^+.$$

For large $n > 0$ these quantities are near to the value of φ at x. For $-n$ one has

$$\mathbb{E}(\varphi | \mathscr{A}_{-n})(x) = \frac{1}{m_+(T^n F(0, T^{-n} x) T^{-n})} \int_{T^n F(0, T^{-n} x) T^{-n}} \varphi(\theta^+ x) \, d\theta^+.$$

Let φ be a C^∞ function. We take a point $x \in T^n W_n^{\delta, \beta}$. We have: $\Delta^+(\beta^n) T^{-n} x \subset \mathscr{Q}_0^\infty(T^{-n} x)$. The inequality (1.5) gives:

$$|\mathbb{E}(\varphi|\mathscr{A}_{-n})(x)| = |\frac{1}{m_+(T^n F(0, T^{-n}x)T^{-n})} \int_{T^n F(0,T^{-n}x)T^{-n}} \varphi(\theta^+ x)\, d\theta^+|$$

$$\leq C(\frac{\mu(\partial U(\rho^n)) + \zeta^{-n}\rho^{Cn}}{\beta^{-nD}} + \beta^{-n}),$$

and, as

$$\mu(\partial U(\rho^{-n})) \leq C\rho^{-n},$$

$$|\mathbb{E}(\varphi|\mathscr{A}_{-n})(x)| \leq C(\frac{\rho^{-n} + \zeta^{-n}\rho^{Cn}}{\beta^{-nD}} + \beta^{-n}).$$

Taking ρ and $\beta > 1$ sufficiently near 1, this shows the existence of $\zeta > 1$ and $C > 0$ such that

$$|\mathbb{E}(\varphi|\mathscr{A}_{-n})(x)| \leq C\Xi^{-n}.$$

As $\mu({}^c W_n^{\delta,\beta}) \leq C\beta^{-n}$, we have:

$$\mathbb{E}(\mathbb{E}(\varphi|\mathscr{A}_{-n})^2) = \int_{T^n W_n^{\delta,\beta}} \mathbb{E}(\varphi|\mathscr{A}_{-n})^2(x)\, d\mu(x)$$

$$+ \int_{T^{nc} W_n^{\delta,\beta}} \mathbb{E}(\varphi|\mathscr{A}_{-n})^2(x)\, d\mu(x) \leq C\Xi^{-n} + C\beta^{-n}\|\varphi\|_\infty^2.$$

This proves the convergence of the first series. The convergence of the second one is trivial.   □

We say that a function φ is η-Hölder continuous if the following quantity is finite:

$$C_\varphi^{(\eta)} = \sup_{y \neq x \in X} \frac{|\varphi(x) - \varphi(y)|}{d(x, y)^\eta}.$$

The preceding theorem is still true if φ is η-Hölder continuous or the characteristic function of a set with smooth boundary (and such a function is never a coboundary). One can prove this by regularizing φ by convolution:

$$x \mapsto \int_G \varphi(g)\chi_n(xg^{-1})\, dg.$$

One has

$$\int_G \varphi(g)\chi_n(xg^{-1})dg = \int_G \varphi(g^{-1}x)\chi_n(g)dg.$$

The first expression shows that the function defined is C^∞, the second allows to control the difference with φ:

$$\left| \varphi(x) - \int_G \varphi(g) \chi_n(xg^{-1}) \, dg \right| \le \int_G \left| \varphi(x) - \varphi(g^{-1}x) \right| \chi_n(g) dg.$$

Now, if $\varphi = \mathbf{1}_F - \mu(F)$, the last expression gives

$$\int_X \left| \varphi(x) - \int_G \varphi(g) \chi_n(xg^{-1}) \, dg \right| d\mu(x) \le C\mu(\partial F(\rho^{-n})).$$

If φ is Hölder continuous or the characteristic function of a set with smooth boundary then it is well approximated by the regularized functions and we apply the preceding reasoning by regularizing not only $\mathbf{1}_U$ but also φ.

1.4.2 Example 6: The Geodesic Flow on a Surface with Constant Curvature of Finite Volume

Let X be the unit tangent bundle of a connected finite volume hyperbolic surface (of constant curvature). We can describe the structure of X as follows: there exists a compact subset X_0 of X such that $X \setminus X_0$ is a finite union of cusps. In the upper-half plane model of the hyperbolic space, a cusp is described as $I \times]c, +\infty[$ with I a compact subset of $\mathbb{R}$.

The preceding technique still works in this new situation. But of course here it is no longer true that every image of a large stable cube in X is well distributed. We know that there are periodic horocycles of length arbitrary small. But we also know that very long periodic horocycles are well distributed in the modular surface. The exponential decay of correlations still holds in the finite volume case. The reasoning used in the compact case can be adapted. We say that F defines a (F, ϵ)-cube at x if

$$[-\epsilon, \epsilon]^2 \times F \to X \ : \ (v, t, w) \longmapsto \theta_v^- g_t \theta_w^+ x$$

is a diffeomorphism on its image in G/Γ, denoted $U_x^{F,\epsilon}$. What shows our computations is that if F defines a (F, ϵ)-cube at x then

$$\left| \frac{1}{m_s(T^n FT^{-n})} \int_{T^n FT^{-n}} \varphi(\theta^+ T^n x) \, d\theta^+ \right|$$

$$\le C\left(\frac{\mu(\partial U_x^{F,\epsilon}(\rho^{-n})) + \zeta^{-n} \rho^{Cn}}{\epsilon^3} + \epsilon \right).$$

We define the partition $\mathscr{Q}_0^\infty$ as in the compact case but starting with a denumerable cover by (F, ϵ)-cubes as large as possible. We adapt the size to the height in the cusps: the diameters of the cubes are of order e^{-n} at height e^n.

Because of the cusps the size of the set of points at a distance less than ϵ of the boundary is different:

$$\mu\{x \in G/\Gamma \ : \ \Delta_s(\epsilon)x \cap \partial\mathscr{Q}(x) \neq \emptyset\} \leq C\epsilon \ln \epsilon^{-1}.$$

This changes a little bit the estimation of the measure of $^c W_n^{\delta,\beta}$: there exists $C > 0$ such that

$$\mu(^c W_n^{\delta,\beta}) \leq Cn\beta^{-n}.$$

But the computations are very similar and leads to the same conclusion.[5]

1.4.3   Example 7: The Diagonal Flows on Compact Quotients of SL(d, ℝ)

Here G is the group $SL(d, \mathbb{R})$ and Γ a cocompact discrete subgroup of G, μ is the probability on G/Γ deduced from the Haar measure on G.

Let $(T_i)_{i=1}^d$ be a decreasing sequence of d positive numbers not all 1 the product of which is 1. Let T be the matrix

$$T = \begin{pmatrix} T_1 & & & & \\ & T_2 & & 0 & \\ & & \ddots & & \\ & 0 & & T_{d-1} & \\ & & & & T_d \end{pmatrix}.$$

The group

$$\{T^t = \begin{pmatrix} T_1^t & & & & \\ & T_2^t & & 0 & \\ & & \ddots & & \\ & 0 & & T_{d-1}^t & \\ & & & & T_d^t \end{pmatrix} \ / \ t \in \mathbb{R}\}$$

---

[5]For more details see [13].

defines a flow on G/Γ, called a diagonal flow, still denoted T^t, that preserves the measure μ

$$T^t : G/\Gamma \longrightarrow G/\Gamma \; : \; x \longmapsto T^t x.$$

We consider the discrete dynamical system $(X, \bar{\mu}, T)$ where $X = SL(d, \mathbb{R})/\Gamma$ and $T = T^1$. The relations

$$\left(TxT^{-1}\right)_{ij} = \frac{T_i x_{ij}}{T_j}$$

permit to identify the stable, unstable and neutral leaves of the diffeomorphism T. Consider the partition of $\{1, \ldots, d\}$ in the sets J_k defined by:

- For every k, for every i, j in J_k, one has $T_i = T_j$.
- For all k, n such that $k < n$, for every i in J_k, every j in J_n, one has $T_i > T_j$.

Let $h_{J_i J_j}$ be a matrix indexed by the set $J_i \times J_j$, Id_{J_i} be the identity matrix indexed by J_i. The unstable leaf of x is the immersed manifold $H_u x$ defined by the group H_u of the matrices:

$$h_u = \begin{pmatrix} Id_{J_1} & h_{J_1 J_2} & \ldots & h_{J_1 J_{l-1}} & h_{J_1 J_l} \\ 0 & Id_{J_2} & \ldots & h_{J_2 J_{l-1}} & h_{J_2 J_l} \\ \vdots & \vdots & \ddots & \vdots & \vdots \\ 0 & 0 & \ldots & Id_{J_{l-1}} & h_{J_{l-1} J_l} \\ 0 & 0 & \ldots & 0 & Id_{J_l} \end{pmatrix}.$$

The stable leaf of x is the immersed manifold $H_s x$ defined by the group H_s of the transposes of elements of H_u. The neutral leaf of x is the immersed manifold $H_e x$ H_e of the matrices:

$$h_e = \begin{pmatrix} h_{J_1 J_1} & 0 & \ldots & 0 \\ 0 & h_{J_2 J_2} & \ldots & 0 \\ \vdots & \vdots & \ddots & \vdots \\ 0 & 0 & \ldots & h_{J_l J_l} \end{pmatrix}.$$

The diffeomorphism T is quasi-hyperbolic.

Starting with a cover of X by a finite set of boxes the sides built on the local decomposition of G as a product of H_u, H_s and H_e, we define $\mathscr{D}$, $\mathscr{D}_0^\infty$, $\mathscr{A}_n$ as in example 5.

Let us fix $\epsilon > 0$ and consider a set $F \subset H_u$ containing $\Delta_u(\epsilon)$ such that the map

$$B_s(Id, \epsilon) \times B_e(Id, \epsilon) \times F \longrightarrow G \; : \; (h_s, h_e, h_u) \longmapsto h_s h_e h_u x$$

is a diffeomorphism on its image. A similar reasoning as in the preceding example gives

$$\left| \frac{1}{m_u(T^n FT^{-n})} \int_{T^n FT^{-n}} \varphi(h_u T^n x)\, dh_u \right|$$

$$\leq C\left(\frac{\mu(\partial U(\rho^{-n})) + \zeta^{-n}\rho^{Cn}}{\epsilon^D} + \epsilon^p \right).$$

To obtain the good distribution of the atoms of $\mathscr{A}_n$, an information on the regularity of their boundaries is needed (we want to bound the term $\mu(\partial U(\rho^{-n}))$). Here we don't have pieces of orbits of an action of $\mathbb{R}^d$ but of the group H_u. It is not abelian but nilpotent. Using its nilpotency one shows that there exist a compact $K \subset \mathbb{R}^{d_s}$ and $M > 0$ such that, for every k, for every x, there exists a set E_x included in K the boundary of which is the union of at most M pieces of algebraic manifolds of degree at most d such that:

$$\mathscr{Q}(x) = exp(E_x)x.$$

Let λ_u the Lebesgue measure on $\mathbb{R}^{d_u}$. Let K be a compact subset of $\mathbb{R}^{d_u}$ and $\epsilon > 0$. There exists a constant C depending uniquely on K such that, for every algebraic hypersurface $S \subset \mathbb{R}^{d_u}$ of degree at most d, one has

$$\lambda_u\{x \in K \ / \ d(x, S) < \epsilon\} \leq C\epsilon,$$

We define $W_n^{\delta,\beta}$ by $(\delta, \beta \in]1, \infty[)$

$$W_n^{\delta,\beta} = \{x \in G/\Gamma \ / \ \forall k \geq 0 \ \Delta_u(\beta^{-n}\delta^{-k})T^{-k}x \subset \mathscr{Q}(T^{-k}x)\}.$$

The group H_u is expanded by the transformation T: there exists $\xi > 1$ such that, for $r > 0$, the set $T^1 \Delta_u(r)T^{-1}$ contains $\Delta_u(\xi r)$.
   If $\delta < \xi$ then, for x in $W_n^{\delta,\beta}$,

$$\Delta_u(\beta^{-n})x \subset \mathscr{Q}_0^\infty(x).$$

The boundaries of the elements of $\mathscr{Q}$ are regular, there exists $C > 0$ such that, for every $\epsilon > 0$, one has

$$\mu\{x \in G/\Gamma \ : \ \Delta_u(\epsilon)x \cap \partial\mathscr{Q}(x) \neq \emptyset\} \leq C\epsilon.$$

The same computation as in the preceding example shows that there exists $C > 0$ such that

$$\mu(^c W_n^{\delta,\beta}) \leq C\beta^{-n}.$$

When x lies in $W_n^{\delta,\beta}$, for k larger than some cn,

$$\mathscr{Q}(x) \subset T^k \mathscr{Q}(T^{-k}x).$$

This implies that the infinite intersection

$$\mathscr{Q}_0^\infty(x) = \mathscr{Q}(x) \cap T \mathscr{Q}(T^{-1}x) \cap T^2 \mathscr{Q}(T^{-2}x) \cap \dots$$

equals the intersection of the cn first sets. This implies that the number of "faces" of $\mathscr{Q}_0^\infty(x)$ is bounded by Cn. As a consequence one gets that there exists a constant $C > 0$ such that, if $x \in W_n^{\delta,\beta}$ then

$$\lambda_u\{y \in \mathscr{Q}_0^\infty(x) \,/\, \Delta_u(\epsilon)x \cap \partial \mathscr{Q}_0^\infty(x) \neq \emptyset\} < Cn\epsilon.$$

This allows us to bound $\mu(\partial U(\rho^{-n}))$ by $Cn\rho^{-n}$ in the expression of the conditional expectation at points of $T^n W_n^{\delta,\beta}$. It insures the convergence of the series.[6]

1.4.4 Examples of Geometrical Applications

The CLT can be used to study the ergodic properties of geodesic flows on some manifolds of infinite volume. We will briefly show it for the surfaces of constant negative curvature that are fibered above a finite volume with $\mathbb{Z}^d$-fibers [12, 23, 45].

Once again we study the time-one map associated. It can be represented as a skew-product over the time one geodesic flow in the finite volume case:

$$T_\varphi : X \times \mathbb{Z}^d \to X \times \mathbb{Z}^d : (x, y) \mapsto (Tx, y + \varphi(x)),$$

where T is the time-one map of the flow defined on the base (of finite volume) and φ is a function with values in $\mathbb{Z}^d$ describing the displacement in the fibers. The iterates of T_φ are given by:

$$T_\varphi^n(x, y) = (T^n x, y + S_n\varphi(x))$$

Traditionally the local limit theorem is used to get recurrence criteria for the cocycle $S_n\varphi$: if the probability $\mathbb{P}(S_n\varphi \in B)$ (where B is a ball) is equivalent to $cn^{-d/2}$ then the cocycle is recurrent for (and only for) $d \leq 2$. For $d = 2$ the (non degenerated) CLT suffices to get the recurrence. We have the following theorem [12] (see also [45]).

---

[6]For more details see [34].

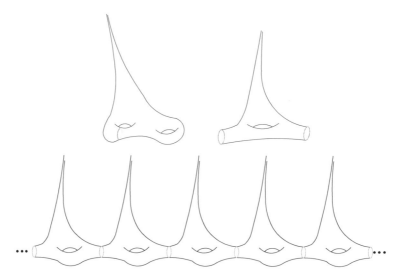

Fig. 1.3 Cutting and gluing a surface along a periodic geodesic to get a $\mathbb{Z}$-cover

Theorem 1.12. *Let (X, T, μ) be a dynamical system and φ a function with values in $\mathbb{R}^2$. If φ satisfies the CLT for the subsequences then the cocycle $S_n\varphi$ is recurrent.*

If moreover (X, T, μ) is a K-system, then recurrence may imply the ergodicity. It is thus possible to deduce that the flow with fibers $\mathbb{Z}^d$ is ergodic if and only if $d \leq 2$ (see [23] for more details).

Let $PSL(2, \mathbb{R})/\Gamma_0$ the unit tangent bundle of a finite volume hyperbolic surface. Cutting the surface along a periodic geodesic or two and gluing together copies along the chosen geodesic(s) define new surfaces of infinite volume (cp. Fig. 1.3) whose unit tangent bundle are given by $PSL(2, \mathbb{R})/\Gamma$ with $\Gamma_0/\Gamma = \mathbb{Z}$ or $\Gamma_0/\Gamma = \mathbb{Z}^2$.

In such a case the function φ describing the displacement in the fibers has finitely many values. If we consider the time-ϵ map of the flow for ϵ sufficiently small, then the values of φ are $\pm$ the generators of $\mathbb{Z}$ or $\mathbb{Z}^2$. Moreover the sets where φ takes its different values are strips around the periodic orbits along which one cuts; these are regular sets. Thus by convolution the theorem of Sect. 1.4.2 is satisfied by φ. So is the CLT along subsequences. Using the result of [12] we deduce that the geodesic flow on $PSL(2, \mathbb{R})/\Gamma$ is recurrent.

What happens for $d \geq 3$? The geodesic flow is now transient. The CLT can be used to precise some aspects of the behaviour of the flow.

More generally, we can study the behaviour of some stationary random walks on $\mathbb{R}^d$. Let (X, T, μ) be an ergodic dynamical system and φ be a measurable function on X with values in $\mathbb{R}^d$, with $d \geq 2$. The ergodic sums $\sum_{k=0}^{n-1} \varphi(T^k x)$ define a vectorial process. When φ is integrable and not centered then the ergodic sums tend a.s. to ∞ in the direction of the mean $\int \varphi \, d\mu$. Consider the case when φ is centered. A question is: in which directions at infinity the ergodic sums are they going? When

φ satisfies a CLT, one can think that the sums behave analogously to a Brownian motion.

Let (B_t) denote the standard Brownian motion in $\mathbb{R}^d$, for $d \geq 2$. If $\mathcal{C}$ is a cone with non empty interior in $\mathbb{R}^d$, the amount of time spent by B_t in $\mathcal{C}$ is

$$\tau_{\mathcal{C}}(t) = \int_0^t \mathbf{1}_{\mathcal{C}}(B_s) \, ds.$$

Theorem 1.13. *Let $\mathcal{C}$ be a cone with non empty interior and non empty exterior and boundary $\partial \mathcal{C}$ of measure 0. We have a.s.:*

$$\limsup_{t \to \infty} \frac{\tau_{\mathcal{C}}(t)}{t} = 1 \quad and \quad \liminf_{t \to \infty} \frac{\tau_{\mathcal{C}}(t)}{t} = 0.$$

Let (X, T, μ) be an ergodic dynamical system and φ be a measurable function on X with values in $\mathbb{R}^d$, $d \geq 2$. We assume that φ is bounded and centered. Let $\mathcal{C}$ be a cone with non empty interior and boundary $\partial \mathcal{C}$ of measure 0.

Let $(W_n)_{n \geq 1}$ be the interpolated piecewise affine process with continuous paths defined for $x \in X$ and $n \geq 1$ by

$$W_n(x, s) = \varphi_k(x) + (ns - k)(\varphi_{k+1}(x) - \varphi_k(x)) \text{ if } s \in [\frac{k}{n}, \frac{k+1}{n}[.$$

This is a process with values in $(\mathcal{C}_d([0, 1]), \|\|_\infty)$ is the space of continuous functions from $[0, 1]$ to $\mathbb{R}^d$.

We say that the invariance principle holds if the stochastic process $(\frac{W_n(x,.)}{\sqrt{n}})_{n \geq 1}$ (defined on the probability space (X, μ) and with values in $\mathcal{C}_d([0, 1])$) converges in distribution to the standard Brownian motion in $\mathbb{R}^d$. The amount of time spent by $W_n(x, s)$ in $\mathcal{C}$ is

$$\tau_{n,\mathcal{C}}(x) = \int_0^1 \mathbf{1}_{\mathcal{C}}(W_n(x, s)) \, ds.$$

We have the following result [14].

Theorem 1.14. *Suppose that (X, T, μ) is ergodic, that the invariance principle is satisfied for a centered function $\varphi : X \to \mathbb{R}^d$ and that $\mathcal{C}$ is a cone with non empty interior, with a complementary with non empty interior and a boundary of Lebesgue measure null. Then, for almost every x,*

$$\limsup_{n \to \infty} \tau_{n,\mathcal{C}}(x) = 1 \text{ and } \liminf_{n \to \infty} \tau_{n,\mathcal{C}}(x) = 0.$$

Let us come back to $\mathbb{Z}^d$-fibered hyperbolic surfaces ($d \geq 3$). For surfaces obtained by gluing copies of one finite volume surface cut along three (or more) periodic geodesic orbits, the multidimensional Donsker invariant principle holds for φ the finite valued function that describes the displacements in the fibers. We deduce from the preceding theorem that almost surely the proportion of the time which the geodesic flow spends in a given cone oscillates infinitely often between 0 and 1.

1.5 Mixing and Equidistribution

1.5.1 Mixing and Directional Regularity

The content of this section has been essentially given in [35]. We take the notations used in example 5. A function φ is said to be η-Hölder continuous on $X = G/\Gamma$ if:

$$C_\varphi^{(\eta)} = \sup_{y \neq x \in X} \frac{|\varphi(x) - \varphi(y)|}{d(x, y)^\eta} < \infty.$$

Proposition 1.5. *Let $F \subseteq \theta^+$ be a set of diameter less than r_0 such that for some $C > 0$ and α, for every $\beta > 0$:*

$$m_u(\partial F(\beta)) \leq C\beta^\alpha.$$

There exist $\xi > 1$ and $C > 0$ such that, for every centered integrable function φ, for every $n \geq 1$ and $x \in G/\Gamma$ one has:

$$\left| \frac{1}{m_u(T^n FT^{-n})} \int_{T^n FT^{-n}} \varphi(\theta_u^+ x) \, d\theta_u^+ \right| \leq \frac{C}{m_u(F)} \left(\|\varphi\|_\infty + C_\varphi^{(\eta)} \right) \xi^{-n}.$$

But the regularity of φ in the direction of θ^+ has no importance here: we compute an integral in this direction. We can precise this inequality, using directional regularities. Let us define

$$C_\varphi^{(\eta,+)} = \sup_{x \in X} \sup_{u \in \mathbb{R}^{d-1}} \frac{|\varphi(x) - \varphi(\theta_u^+ x)|}{\|u\|^\eta},$$

and analogously

$$C_\varphi^{(\eta,0,-)} = \sup_{x \in X} \sup_{v \in \mathbb{R}^{d-1}, k \in K} \frac{|\varphi(x) - \varphi(\theta_v^+ k x)|}{d(x, \theta_v^+ k x)^\eta}.$$

Proposition 1.6. *Let $F \subseteq \theta^+$ be a set of diameter less than r_0 such that for some $C > 0$ and α, for every $\beta > 0$:*

$$m_u(\partial F(\beta)) \leq C\beta^\alpha.$$

There exist $\xi > 1$ and $C > 0$ such that, for every centered integrable function φ, for every $n \geq 1$ and $x \in G/\Gamma$ one has:

$$\left| \frac{1}{m_u(T^n F T^{-n})} \int_{T^n F T^{-n}} \varphi(\theta_u^+ x) \, d\theta_u^+ \right| \leq \frac{C}{m_u(F)} \left(\|\varphi\|_\infty + C_\varphi^{(\eta,0,-)} \right) \xi^{-n}.$$

Proof. Consider a function φ supported in $P = \theta_{[-r_0,r_0]^{d-1}}^+ B_K(r_0) \theta_{[-r_0,r_0]^{d-1}}^- x$. We regularize φ in direction θ^+: let f be a C^∞ function supported in $\theta_{[-r_0,r_0]^{d-1}}^+$, non negative, with integral 1 ($C_f^{(1)}$ is the Lipschitz constant of f). Let us define ψ by:

$$\psi(\theta_u^+ k \theta_s^- x) = f(\theta_u^+) \int_{\Delta^+(r_0)} \varphi(\theta_{u'}^+ k \theta_s^- x) \, d\theta_{u'}^+,$$

if $(\theta_u^+, k, \theta_s^-) \in \Delta^+(r_0) \times B_K(r_0) \times \Delta^-(r_0)$, and $\psi(y) = 0$ if y does not lie in P.

Lemma 1.3. *The functions φ and ψ have the same integral and*

$$C_\psi^{(\eta)} \leq C \left(\|\varphi\|_\infty + C_\varphi^{(\eta,0,-)} \right).$$

Proof. The first point follows from the theorem of Fubini. Let us show the bounding. Take $(\theta_{u_1}^+, k_1, \theta_{s_1}^-)$ and $(\theta_{u_2}^+, k_2, \theta_{s_2}^-)$ two points in $\Delta^+(r_0) \times B_K(r_0) \times \Delta^-(r_0)$ one has:

$$\left| \psi(\theta_{u_1}^+ k_1 \theta_{s_1}^- x) - \psi(\theta_{u_2}^+ k_2 \theta_{s_2}^- x) \right|$$

$$= \left| f(\theta_{u_1}^+) \int_{\Delta^+(r_0)} \varphi(\theta_{u'}^+ k_1 \theta_{s_1}^- x) \, d\theta_{u'}^+ - f(\theta_{u_2}^+) \int_{\Delta^+(r_0)} \varphi(\theta_{u'}^+ k_2 \theta_{s_2}^- x) \, d\theta_{u'}^+ \right|$$

$$\leq \|\varphi\|_\infty |f(\theta_{u_1}^+) - f(\theta_{u_2}^+)| m_u(\Delta^+(r_0))$$

$$+ \|f\|_\infty \left| \int_{\Delta^+(r_0)} \varphi(\theta_{u'}^+ k_1 \theta_{s_1}^- x) \, d\theta_{u'}^+ - \int_{\Delta^+(r_0)} \varphi(\theta_{u'}^+ k_2 \theta_{s_2}^- x) \, d\theta_{u'}^+ \right|.$$

Let $\Pi_{0,-}$ denote the holonomy map along the neutral/stable direction $K\theta^-$ from $\Delta^+(2r_0) k_1 \theta_{s_1}^- x$ to $\Delta^+(3r_0) k_2 \theta_{s_2}^- x$. This is a local projection (see Fig. 1.4) defined by

$$\Pi_{0,-}(\theta_u^+ k_1 \theta_{s_1}^-) = \theta_{u'}^+ k_2 \theta_{s_2}^-$$

if there exists (k, θ^-) in $B_K(2r_0) \times \Delta^-(2r_0)$ such that

$$k \theta^- \theta_u^+ k_1 \theta_{s_1}^- = \theta_{u'}^+ k_2 \theta_{s_2}^-.$$

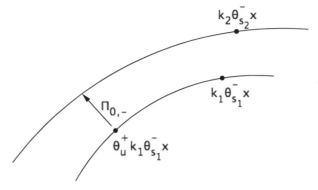

Fig. 1.4 Local projection from one unstable leaf onto another along the weak stable leaves

This holonomy has the following regularity properties:

$$d(\Pi_{0,-}(\theta_u^+ k_1 \theta_{s_1}^- x), \theta_u^+ k_1 \theta_{s_1}^- x) \le Cd(k_1 \theta_{s_1}^- x, k_2 \theta_{s_2}^- x),$$

$$\left| J_{\Pi_{0,-}}(\theta_u^+ k_1 \theta_1^- x) - 1 \right| \le Cd(k_1 \theta_{s_1}^- x, k_2 \theta_{s_2}^- x),$$

where $J_{\Pi_{0,-}}$ is the jacobian determinant of $\Pi_{0,-}$.
We have:

$$\left| \int_{\Delta^+(r_0)} \varphi(\theta_{u'}^+ k_1 \theta_{s_1}^- x) \, d\theta_{u'}^+ - \int_{\Delta^+(r_0)} \varphi(\theta_{u'}^+ k_2 \theta_{s_2}^- x) \, d\theta_{u'}^+ \right|$$

$$= \left| \int_{\Delta^+(r_0)} \varphi(\theta_{u'}^+ k_1 \theta_{s_1}^- x) \, d\theta_{u'}^+ - \int_{\Pi_{0,-}^{-1}(\Delta^+(r_0))} (\varphi \circ \Pi_{0,-}.J_{\Pi_{0,-}})(\theta_{u'}^+ k_1 \theta_1^- x) d\theta_{u'}^+ \right|$$

$$\le \int_{\Delta^+(2r_0)} \left| \varphi(\theta_{u'}^+ k_1 \theta_{s_1}^- x) - \varphi(\Pi_{0,-}(\theta_{u'}^+ k_1 \theta_1^-)x) \right| \, d\theta_{u'}^+$$

$$+ \int_{\Delta^+(2r_0)} \left| \varphi(\Pi_{0,-}(\theta_{u'}^+ k_1 \theta_1^-)x) \right| . \left| J_{\Pi_{0,-}}(\theta_u^+ k_1 \theta_1^- x) - 1 \right| \, d\theta_{u'}^+$$

$$\le Cm_u(\Delta^+(2r_0))C_\varphi^{(\eta,0,-)} d(k_1 \theta_{s_1}^- x, k_2 \theta_{s_2}^- x)^\eta$$

$$+ C\|\varphi\|_\infty m_u(\Delta^+(2r_0))d(k_1 \theta_{s_1}^- x, k_2 \theta_{s_2}^- x)$$

$$\le C(\|\varphi\|_\infty + C_\varphi^{(\eta,0,-)})d(k_1 \theta_{s_1}^- x, k_2 \theta_{s_2}^- x)^\eta.$$

We thus have:

$$\left| \psi(\theta_{u_1}^+ k_1 \theta_{s_1}^- x) - \psi(\theta_{u_2}^+ k_2 \theta_{s_2}^- x) \right|$$

$$\le \|f\|_\infty C \left(\|\varphi\|_\infty + C_\varphi^{(\eta,0,-)} \right) d \left(k_1 \theta_{s_1}^-, k_2 \theta_{s_2}^- \right)^\eta + C\|\varphi\|_\infty d(\theta_{u_1}^+, \theta_{u_2}^+)$$

$$\le C \left(\|\varphi\|_\infty + C_\varphi^{(\eta,0,-)} \right) d \left(\theta_{u_1}^+ k_1 \theta_{s_1}^-, \theta_{u_2}^+ k_2 \theta_{s_2}^- \right)^\eta, \qquad\qquad \square$$

Let us come back to the proof of Proposition 1.6. The function ψ being η-Hölder continuous, we apply to it Proposition 1.5. For every $n \geq 1$, every x, we have:

$$\left| \frac{1}{m_u(T^n FT^{-n})} \int_{T^n FT^{-n}} \psi(\theta_u^+ x) \, d\theta_u^+ \right| \leq \frac{C \left(\|\psi\|_\infty + C_\psi^{(\eta)} \right) \xi^{-n}}{m_u(F)}$$

and, because of the lemma:

$$\left| \frac{1}{m_u(T^n FT^{-n})} \int_{T^n FT^{-n}} \psi(\theta_u^+ x) \, d\theta_u^+ \right| \leq \frac{C \left(\|\varphi\|_\infty + C_\varphi^{(\eta,0,-)} \right) \xi^{-n}}{m_u(F)}.$$

We now just have to estimate the difference between the two integrals:

$$\frac{1}{m_u(T^n FT^{-n})} \int_{T^n FT^{-n}} \psi(\theta_u^+ x) \, d\theta_u^+ \quad \text{and} \quad \frac{1}{m_u(T^n FT^{-n})} \int_{T^n FT^{-n}} \varphi(\theta_u^+ x) \, d\theta_u^+.$$

These integrals are sums of integrals of φ and ψ on the connected components of the intersection of $P = \Delta^+(r_0) B_K(r_0) \Delta^-(r_0) x$ with $T^n FT^{-n}$. The integrals on connected components may be different only on pieces containing a point of the boundary of $T^n FT^{-n} x$ (on entire "slices" the integrals of φ and ψ are equal by definition of ψ). We thus have:

$$\left| \frac{1}{m_u(T^n FT^{-n})} \int_{T^n FT^{-n}} (\psi(\theta_u^+ x) - \varphi(\theta_u^+ x)) \, d\theta_u^+ \right| \leq 2\|\varphi\|_\infty \frac{m_u(\partial T^n FT^{-n}(r_0))}{m_u(T^n FT^{-n})}.$$

Let θ_u^+ be a point of $T^n FT^{-n}$ at a distance less then r_0 from the boundary of $T^n FT^{-n}$; there exists $\theta_{u'}^+ \in \partial T^n FT^{-n}$ such that: $d(\theta_u^+, \theta_{u'}^+) \leq r_0$. Thus we have $d(T^{-n}\theta_u^+ T^n, T^{-n}\theta_{u'}^+ T^n) \leq Cr_0 e^{-n}$, that is $T^{-n}\theta_{u'}^+ T^n$ is in $\partial F(Cr_0 e^{-n})$. We deduce that $\partial T^n FT^{-n}(r_0)$ is included in $T^n \partial F(Cr_0 e^{-n})T^{-n}$. The action of T on θ^+ by conjugacy is linear, consequently:

$$\frac{m_u(\partial T^n FT^{-n}(r_0))}{m_u(T^n FT^{-n})} \leq \frac{m_u(T^n \partial F(Cr_0 e^{-n})T^{-n})}{m_u(T^n FT^{-n})}$$

$$\leq \frac{m_u(\partial F(Cr_0 e^{-n}))}{m_u(F)},$$

so that

$$\left| \frac{1}{m_u(T^n FT^{-n})} \int_{T^n FT^{-n}} (\psi(\theta_u^+ x) - \varphi(\theta_u^+ x)) \, d\theta_u^+ \right| \leq C \frac{\|\varphi\|_\infty e^{-n}}{m_u(F)},$$

and

$$\left| \frac{1}{m_u(T^n F T^{-n})} \int_{T^n F T^{-n}} \varphi(\theta_u^+ x) \, d\theta_u^+ \right| \leq C \frac{\left(\|\varphi\|_\infty + C_\varphi^{(\eta,0,-)} \right) \xi^{-n}}{m_u(F)}.$$

We get rid of the support condition on φ through a partition of the unity associated to a finite cover of G/Γ by sets like $\Delta^+(r_0) B_K(r_0) \Delta^-(r_0) y$. $\square$

From this result it is possible to get back to mixing and to obtain a new statement, an anisotropic mixing property.

Theorem 1.15. *Let φ and ψ be two centered Hölder-continuous functions on G/Γ. There exist two numbers $\zeta > 1$ and $C > 0$ and such that, for every n, one has:*

$$|\langle \varphi, \psi \circ T^{-n} \rangle| \leq C \left(\|\varphi\|_\infty + C_\varphi^{(\eta,0,-)} \right) \left(\|\psi\|_\infty + C_\psi^{(\eta,+)} \right) \zeta^{-|n|}. \qquad (1.6)$$

Proof. Consider a sequence of σ-algebras, $(\mathscr{A}_n)$ the atoms of which are regular pieces of unstable leaves, and satisfying $\mathscr{A}_n = T^{-n} \mathscr{A}_0$ (we do not need a filtration, so that the m_u size of the atoms of $\mathscr{A}_0$ can be chosen bounded from below). We have:

$$
\begin{aligned}
|\langle \varphi, T^{-n} \psi \rangle| &\leq |\langle \varphi, T^{-n}(\psi - \mathbb{E}(\psi|\mathscr{A}_{n/2})) \rangle| + |\langle \varphi, T^{-n} \mathbb{E}(\psi|\mathscr{A}_{n/2}) \rangle| \\
&\leq C C_\psi^{(\eta,+)} \|\varphi\|_\infty \zeta^{-n} + |\langle \varphi, \mathbb{E}(T^{-n}\psi|\mathscr{A}_{-n/2}) \rangle| \\
&\leq C C_\psi^{(\eta,+)} \|\varphi\|_\infty \zeta^{-n} + |\langle \mathbb{E}(\varphi|\mathscr{A}_{-n/2}), \mathbb{E}(T^{-n}\psi|\mathscr{A}_{-n/2}) \rangle| \\
&\leq C C_\psi^{(\eta,+)} \|\varphi\|_\infty \zeta^{-n} + C \|\psi\|_\infty \left(\|\varphi\|_\infty + C_\varphi^{(\eta,0,-)} \right) \xi^{-n/2}. \qquad \square
\end{aligned}
$$

1.5.2   Example 8: Composing Different Transformations

We will show that the strong mixing property obtained in the preceding section implies the CLT. This will be done through simple computations introduced by Jan in [29]. The fact that we do not need to construct a filtration simplify considerably the construction of the $\mathscr{A}_n$. But the study of the characteristic function of the normalized sums is more involved than in the martingale case. However Jan's method is elementary and flexible. To illustrate this, we will apply it to a case where martingales could be hard to use and moreover a non stationary case.

The discrete time in Proposition 1.6 is of no importance. The same reasoning leads to the following proposition.

Proposition 1.7. *Let $\mathscr{A}$ be a σ-algebra the atoms of which are regular pieces of orbits of θ^+ containing cubes of the form $\theta_{[-\epsilon,\epsilon]^{d-1}}^+$ (with a uniform ϵ). There exists $C > 0$, $\xi > 1$ such that for any η-Hölder-continuous function φ on X, for every*

$t > 0$, one has

$$|\varphi - \mathbb{E}\left[\varphi \,|g_{-t}\mathscr{A}\right]| \leq C C_\varphi^{(\eta,+)}\xi^{-t} \tag{1.7}$$

and

$$\left\| \mathbb{E}\left[\varphi \,|g_t\mathscr{A}\right] - \int_\Omega \varphi \, dv \right\|_\infty \leq C \left(\|\varphi\|_\infty + C_\varphi^{(\eta,0,-)} \right) \xi^{-t}. \tag{1.8}$$

Now let $(t_k)_{k \geq 1}$ be a sequence of positive times bounded from below (for some $\delta > 0$, $t_k \geq \delta$ for all k) and denote $s_k = \sum_{j=1}^k t_j$. For φ and ψ two η-Hölder-continuous centered functions.

$$|\langle \psi \circ g_{-s_k}, \varphi \rangle| = |\langle (\psi - \mathbb{E}\left[\psi \,|g_{-s_k/2}\mathscr{A}\right]) \circ g_{-s_k}, \varphi \rangle + \langle \mathbb{E}\left[\psi \,|g_{-s_k/2}\mathscr{A}\right] \circ g_{-s_k}, \varphi \rangle|$$
$$\leq C C_\psi^{(\eta,+)}\|\varphi\|_\infty \xi^{-s_k/2} + \langle \mathbb{E}\left[\psi \,|g_{-s_k/2}\mathscr{A}\right] \circ g_{-s_k}, \varphi \rangle|.$$

But

$$\mathbb{E}\left[\psi \,|g_{-s_k/2}\mathscr{A}\right] \circ g_{-s_k} = \mathbb{E}\left[\psi \circ g_{-s_k} \,|g_{s_k/2}\mathscr{A}\right],$$

so that we get

$$|\langle \psi \circ g_{-s_k}, \varphi \rangle| \leq C C_\psi^{(\eta,+)}\|\varphi\|_\infty \xi^{-s_k/2} + \langle \mathbb{E}\left[\psi \circ g_{-s_k} \,|g_{s_k/2}\mathscr{A}\right], \mathbb{E}\left[\varphi \,|g_{s_k/2}\mathscr{A}\right] \rangle|$$

and, because φ is centered,

$$\left\| \mathbb{E}\left[\varphi \,|g_{s_k/2}\mathscr{A}\right] \right\|_\infty \leq C \left(\|\varphi\|_\infty + C_\varphi^{(\eta,0,-)} \right) \xi^{-s_k/2}.$$

Hence one has

$$|\langle \psi \circ g_{-s_k}, \varphi \rangle| \leq C \left(C_\psi^{(\eta,+)}\|\varphi\|_\infty + \|\psi\|_\infty \|\varphi\|_\infty + C_\varphi^{(\eta,0,-)}\|\psi\|_\infty \right) \xi^{-k\delta/2}. \tag{1.9}$$

From this inequality we will deduce the following theorem using Jan's method.

Theorem 1.16. *Let $(t_k)_{k \geq 1}$ be a sequence of positive times bounded from below and denote $s_k = \sum_{j=1}^k t_j$. Let φ a Hölder-continuous function on G/Γ. If the variances of the variables*

$$\frac{1}{\sqrt{n}} \sum_{k=1}^n \varphi \circ g_{s_k}$$

have a positive limit these variables converge in distribution toward a gaussian law.

Proof. Suppose that the following limit exists

$$\sigma^2(\varphi) = \lim_{n \to \infty} \frac{1}{n} \mu\left(\left(\sum_{k=1}^{n} \varphi \circ g_{s_k}\right)^2\right).$$

Let s_l^k be the sum $s_l^k = \sum_{j=l}^{k} t_j$. The mixing property of the proposition (1.9) implies that the quantity $\sigma_\ell^2(\varphi)$ is well defined:

$$\sigma_\ell^2(\varphi) = \mathbb{E}(\varphi^2) + 2 \sum_{k=\ell+1}^{\infty} \mathbb{E}(\varphi \circ g_{s_k} \varphi \circ g_{s_\ell})$$

and that the sequence $(\sigma_\ell^2(\varphi))_\ell$ is bounded. One proves that $\sigma^2(\varphi)$ exists if $(\sigma_\ell^2(\varphi))_\ell$ converges in the sense of Cesaro:

$$\sigma^2(\varphi) = \lim_{n \to \infty} \frac{1}{n} \sum_{\ell=1}^{n} \sigma_\ell^2(\varphi).$$

Let us consider a probability space $(\Omega', \mathbb{P}')$ containing (X, μ) and a sequence (U_k) of bounded independent random variables with variances $\sigma_k^2(\varphi)$ defined on Ω', and independent from the variables $X_\ell = T^\ell \varphi$, of distribution $1/2(\delta_{-\sigma_k(\varphi)} + \delta_{\sigma_k(\varphi)})$. The CLT (with Lyapounov condition) holds for (U_ℓ):

$$\frac{1}{n^{1/2}} \sum U_k \to_{\mathscr{L}} \mathscr{N}(0, \sigma^2).$$

This is a consequence of the fact that the quantity

$$|\mathbb{E}(\exp(\frac{it}{n^{1/2}} \sum_{\ell=1}^{n} U_\ell)) - \exp(-\frac{1}{2}\sigma(\varphi)^2 t^2)|$$

$$= |\prod_{\ell=1}^{n} \cos(\frac{1}{n^{1/2}}\sigma_\ell(\varphi)t) - \exp(-\frac{1}{2}\sigma(\varphi)^2 t^2)|$$

$$= |\prod_{\ell=1}^{n}(1 - \frac{\sigma_\ell^2(\varphi)t^2}{2n} + O(\frac{1}{n^{3/2}})) - \exp(-\frac{1}{2}\sigma(\varphi)^2 t^2)|$$

tends to 0 as n goes to infinity.

We will now study the difference

$$\mathbb{E}(\exp(\frac{it}{n^{1/2}} \sum_{k=1}^{n} \varphi \circ g_{s_k})) - \mathbb{E}(\exp(\frac{it}{n^{1/2}} \sum_{\ell=1}^{n} U_\ell)).$$

We will use the following notations

$$B_{\ell,n} = \exp(\frac{it}{n^{1/2}}\varphi \circ g_{s_\ell}), \quad C_{\ell,n} = \exp(\frac{it}{n^{1/2}}U_\ell).$$

We have

$$\exp\left(\frac{it}{n^{1/2}}\sum_{\ell=1}^{n}\varphi \circ g_{s_\ell}\right) - \exp\left(\frac{it}{n^{1/2}}\sum_{\ell=1}^{n}U_\ell\right) = \prod_{\ell=1}^{n}B_{\ell,n} - \prod_{\ell=1}^{n}C_{\ell,n}, \quad (1.10)$$

and

$$\prod_{\ell=1}^{n}B_{\ell,n} - \prod_{\ell=1}^{n}C_{\ell,n} = \sum_{\ell=1}^{n}(\prod_{k=1}^{\ell-1}C_{k,n})\underbrace{(B_{\ell,n} - C_{\ell,n})(\prod_{k=\ell+1}^{n}B_{k,n})}_{\Delta_\ell},$$

where a product on the empty set is taken equal to 1.

The variables $\Delta_\ell = (B_{\ell,n} - C_{\ell,n})\prod_{k=\ell+1}^{n}B_{\ell,n}$ and $\prod_{k=0}^{\ell-1}C_{\ell,n}$ are independent. We are going to show that most of the n terms $|\mathbb{E}(\Delta_\ell)|$ are bounded by a constant $Cn^{-3/2}\ln n$. It will imply the result.

Let us consider a sequence $(\chi(n))$ defined later (of order $\ln n$). When $\ell + 3\chi(n) + 1 < n$, we decompose the product Δ_ℓ into blocks:

$$\Delta_\ell = \underbrace{(B_{\ell,n} - C_{\ell,n})\prod_{k=\ell+1}^{\ell+\chi(n)}B_{k,n}}_{\mathscr{A}}\underbrace{\prod_{k=\ell+\chi(n)+1}^{\ell+2\chi(n)}B_{k,n}}_{\mathscr{B}}\underbrace{\prod_{k=\ell+2\chi(n)+1}^{\ell+3\chi(n)}B_{k,n}}_{\mathscr{C}}\underbrace{\prod_{k=\ell+3\chi(n)+1}^{n}B_{k,n}}_{\mathscr{D}}.$$

We can write

$$\mathbb{E}(\Delta_\ell) = \mathbb{E}(\mathscr{A}\mathscr{B}\mathscr{C}\mathscr{D}) \qquad (1.11)$$

$$= \mathbb{E}(\mathscr{A}(\mathscr{B} - 1)(\mathscr{C} - 1)\mathscr{D}) + \mathbb{E}(\mathscr{A}\mathscr{B}\mathscr{D}) + \mathbb{E}(\mathscr{A}\mathscr{C}\mathscr{D}) - \mathbb{E}(\mathscr{A}\mathscr{D}).$$

$$(1.12)$$

The mean value theorem implies that $\mathscr{A}$ is bounded by

$$Ctn^{-1/2}(\|\varphi\|_\infty + \|U_\ell\|_\infty),$$

for a constant C, and $(\mathscr{B} - 1)$, $(\mathscr{C} - 1)$ are both bounded by

$$\frac{2t}{n^{1/2}} \sum_{k=\ell+\chi(n)+1}^{\ell+2\chi(n)} |\varphi \circ g_{s_k}|.$$

Hence $\mathbb{E}(\mathscr{A}(\mathscr{B} - 1)(\mathscr{C} - 1)\mathscr{D}) \leq C \|\varphi\|_\infty^3 \frac{t^3}{n^{3/2}} \chi(n)^2$.

We will not retain the dependence in t, nor $\|\varphi\|_\infty$ for our computations. For example we just write

$$\mathbb{E}(\mathscr{A}(\mathscr{B} - 1)(\mathscr{C} - 1)\mathscr{D}) \leq C \frac{1}{n^{3/2}} \chi(n)^2. \tag{1.13}$$

We bound the three other terms in the same following way.
Consider for example: $\mathbb{E}(\mathscr{A}\mathscr{B}\mathscr{D}) = Cov(\mathscr{A}\mathscr{B}, \mathscr{D}) + \mathbb{E}(\mathscr{A}\mathscr{B})\mathbb{E}(\mathscr{D})$.
We have:

$$
\begin{aligned}
Cov(\mathscr{A}\mathscr{B}, \mathscr{D}) &= Cov((B_{\ell,n} - C_{\ell,n}) \prod_{k=\ell+1}^{\ell+\chi(n)} B_{k,n}, \prod_{k=\ell+3\chi(n)+1}^{n-1} B_{k,n}) \\
&= Cov(\prod_{k=\ell}^{\ell+\chi(n)} B_{k,n}, \prod_{k=\ell+3\chi(n)+1}^{n-1} B_{k,n}) \\
&= Cov\left(\prod_{k=\ell}^{\ell+\chi(n)} \exp(\frac{it}{n^{1/2}}\varphi) \circ g_{s_k}, \prod_{k=\ell+3\chi(n)+1}^{n-1} \exp(\frac{it}{n^{1/2}}\varphi) \circ g_{s_k}\right).
\end{aligned}
$$

Let us compose both terms of the preceding expressions by $g_{s_{\ell+\chi(n)}}$ and denote by $\Phi(n, \ell, t)$ and $\Psi(n, \ell, t)$ the functions defined by

$$\Phi(n, \ell, t) = \prod_{k=\ell}^{\ell+\chi(n)} \exp(\frac{it}{n^{1/2}}\varphi) \circ g_{[s_k - s_{\ell+\chi(n)}]}$$

and

$$\Psi(n, \ell, t) = \prod_{k=\ell+3\chi(n)+1}^{n-1} \exp(\frac{it}{n^{1/2}}\varphi) \circ g_{[s_k - s_{\ell+3\chi(n)}]}.$$

All the times appearing in the products defining $\Phi(n, \ell, t)$ and $\Psi(n, \ell, t)$ are respectively negative and positive and we have

$$Cov(\mathscr{A}\mathscr{B}, \mathscr{D}) = Cov(\Psi(n, \ell, t) \circ g_{[s_{\ell+3\chi(n)} - s_{\ell+\chi(n)}]}, \Phi(n, \ell, t)).$$

By hypothesis $s_{\ell+3\chi(n)} - s_{\ell+\chi(n)}$ is larger than $2\chi(n)\delta$. Thus applying (1.9) we get that the covariance $Cov(\mathscr{A}\mathscr{B}, \mathscr{D})$ is bounded (the exponential functions are bounded by 1) by

$$C\left(1 + C^{(\eta,+)}_{\Phi(n,\ell,t)} + C^{(\eta,0,-)}_{\psi(n,\ell,t)}\right)\xi^{-\chi(n)\delta}.$$

But

$$C^{(\eta,+)}_{\Phi(n,\ell,t)} \leq \sum_{k=\ell}^{\ell+\chi(n)} C^{(\eta,+)}_{\exp(\frac{it}{n^{1/2}}\varphi)\circ g_{[s_k - s_{\ell+\chi(n)}]}} \leq (\chi(n)+1)C^{(\eta,+)}_{\exp(\frac{it}{n^{1/2}}\varphi)}$$

because, for $s < 0$, $v \in \mathbb{R}^{d-1}$ and $k \in K$, $d(g_s\theta^+ kx, g_s x) \leq d(\theta^+ kx, x)$. Similarly

$$C^{(\eta,0,-)}_{\psi(n,\ell,t)} \leq nC^{(\eta,0,-)}_{\exp(\frac{it}{n^{1/2}}\varphi)},$$

so that

$$Cov(\mathscr{A}\mathscr{B}, \mathscr{D}) \leq Cn\zeta^{-\chi(n)}. \tag{1.14}$$

Now we study

$$\mathscr{A}\mathscr{B} = (B_{\ell,n} - C_{\ell,n})\prod_{k=\ell+1}^{\ell+2\chi(n)} B_{k,n}$$

$$= \left(\exp\frac{it\varphi \circ g_{s_\ell}}{n^{1/2}} - \exp\frac{itU_\ell}{n^{1/2}}\right)\exp\left(itn^{-1/2}\sum_{k=\ell+1}^{\ell+2\chi(n)}\varphi \circ g_{s_k}\right).$$

From the Taylor expansion of the two other terms at order 2 and 1 we deduce the equalities

$$\exp\frac{it\varphi \circ g_{s_\ell}}{n^{1/2}} - \exp\frac{itU_\ell}{n^{1/2}} = \frac{it}{n^{1/2}}(\varphi \circ g_{s_\ell} - U_\ell) - \frac{1}{2n}(\varphi \circ g_{s_\ell} - U_\ell)^2 + D_1,$$

with $D_1 \leq C\frac{1}{n^{3/2}}$,

$$\exp\left(itn^{-1/2}\sum_{k=\ell+1}^{\ell+2\chi(n)}\varphi \circ g_{s_k}\right) = 1 + itn^{-1/2}\sum_{k=\ell+1}^{\ell+2\chi(n)}\varphi \circ g_{s_k} + D_2,$$

with $D_2 \leq C \frac{\chi(n)^2}{n}$, and

$$\mathscr{A}\mathscr{B} = \frac{it}{n^{1/2}}(\varphi \circ g_{s_\ell} - U_\ell) - \frac{t^2}{2n}(\varphi^2 \circ g_{s_\ell} - U_\ell^2)$$

$$-\frac{t^2}{n}\sum_{k=\ell+1}^{\ell+2\chi(n)} \varphi \circ g_{s_k}\, \varphi \circ g_{s_\ell} + \frac{t^2}{n}U_\ell \sum_{k=\ell+1}^{\ell+2\chi(n)} \varphi \circ g_{s_k} + D,$$

with $D \leq C \frac{\chi(n)^2}{n^{3/2}} + \frac{\chi(n)^3}{n^2} + \frac{\chi(n)^4}{n^{5/2}}$. By taking the expectation, we obtain:

$$|\mathbb{E}(\mathscr{A}\mathscr{B})|$$

$$\leq \frac{t^2}{2n}\left(\mathbb{E}(U_\ell^2) - \left(\mathbb{E}(g_{s_\ell}\varphi^2) + 2\sum_{k=\ell+1}^{\ell+2\chi(n)} \mathbb{E}(g_{s_k}\varphi\, g_{s_\ell}\varphi)\right)\right) + C\frac{\chi(n)^2}{n^{3/2}}. \quad (1.15)$$

But from the definition of U_ℓ, we have

$$\mathbb{E}(U_\ell^2) = \sigma_\ell^2(\varphi) = E(\varphi^2 \circ g_{s_\ell}) + 2\sum_{k=\ell+1}^{\infty} \mathbb{E}(\varphi \circ g_{s_k}\, \varphi \circ g_{s_\ell})$$

$$= E(\varphi^2 \circ g_{s_\ell}) + 2\sum_{k=\ell+1}^{\ell+2\chi(n)} \mathbb{E}(\varphi \circ g_{s_k}\, \varphi \circ g_{s_\ell})$$

$$+2\sum_{k=\ell+2\chi(n)+1}^{\infty} \mathbb{E}(\varphi \circ g_{s_k}\, \varphi \circ g_{s_\ell}).$$

By replacing $\mathbb{E}(U_\ell^2)$ by this expression in (1.15), we obtain

$$|\mathbb{E}(\mathscr{A}\mathscr{B})| \leq C\left(\left|\frac{t^2}{2n}\sum_{k=\ell+2\chi(n)+1}^{\infty} \mathbb{E}(\varphi \circ g_{s_k}\, \varphi \circ g_{s_\ell})\right| + \chi(n)^2 n^{-3/2}\right),$$

and, as the general term of the series tends exponentially fast to 0,

$$|\mathbb{E}(\mathscr{A}\mathscr{B})| \leq C\left(\frac{\zeta^{-\chi(n)}}{n} + \chi(n)^2 n^{-3/2}\right). \quad (1.16)$$

Since $|\mathbb{E}(\mathscr{D})| \leq 1$, (1.14) and (1.16) imply

$$\mathbb{E}(\Delta_\ell) \leq C\left(\frac{\zeta^{-\chi(n)}}{n} + \chi(n)^2 n^{-3/2} + n\zeta^{-\chi(n)}\right).$$

Now we can bound (1.10):

$$|\mathbb{E}(\prod_0^{n-1} B_{\ell,n} - \prod_0^{n-1} C_{\ell,n})| = |\sum_{\ell=0}^{n} \mathbb{E}(\prod_{k=0}^{\ell-1} C_{k,n})\mathbb{E}(\Delta_\ell)| \leq \sum_{\ell=0}^{n} |\mathbb{E}(\Delta_\ell)|$$

$$\leq \sum_{\ell=0}^{n-3\chi(n)-1} C(\frac{\zeta^{-\chi(n)}}{n} + \chi(n)^2 n^{-3/2} + n\zeta^{-\chi(n)})$$

$$+ \sum_{\ell=n-3\chi(n)}^{n} \mathbb{E}(\Delta_\ell).$$

The mean value theorem implies $\mathbb{E}(\Delta_\ell) \leq Cn^{-1/2}$. If we take $\chi(n) = D \ln n$ with D sufficiently large, then

$$\mathbb{E}(\exp(\frac{it}{n^{1/2}} \sum_{k=1}^{n} \varphi \circ g_{s_k})) - \mathbb{E}(\exp(\frac{it}{n^{1/2}} \sum_{\ell=1}^{n} U_\ell)) = |\mathbb{E}(\prod_0^{n} B_{\ell,n} - \prod_0^{n} C_{\ell,n})|$$

$$\leq C \frac{\ln^2(n)}{n^{1/2}}. \qquad \square$$

1.6 Some General References

- On convergence in distribution: [4].
- On representation theory: [6–8, 28].
- On hyperbolic geometry: [15, 20, 41].

Acknowledgements These notes are an extended version of the lectures given at the INdAM in May 2013. Less details were presented on the blackboard. It was a great pleasure to work in Rome. I thank a lot the organizers of the workshop: Françoise Dal'bo, Marc Peigné, and, above all, Andrea Sambusetti for his very kind welcome in Rome. I also thank Bachir Bekka for hints on representation theory (in Rennes).

References

1. D.V. Anosov, Geodesic flows on closed Riemannian manifolds of negative curvature. Trudy Mat. Inst. Steklov. **90**, 209 pp. (1967)
2. R.L. Adler, B. Weiss, Entropy, a complete metric invariant for automorphisms of the torus. Proc. Natl. Acad. Sci. U.S.A. **57**, 1573–1576 (1967)
3. M.B. Bekka, On uniqueness of invariant means. Proc. AMS **126**(2), 507–514 (1998)
4. P. Billingsley, *Convergence of Probability Measures* (Wiley, New York, 1999)
5. M. Blank, G. Keller, C. Liverani, Ruelle-Perron-Frobenius spectrum for Anosov maps. Nonlinearity **15**(6), 1905–1973 (2002)

6. N. Bourbaki, *Éléments de mathématique. Fascicule XXIX. Livre VI: Intégration. Chapitre 7: Mesure de Haar. Chapitre 8: Convolution et représentations.* Actualités Scientifiques et Industrielles, No. 1306 (Hermann, Paris, 1963)
7. N. Bourbaki, *Éléments de mathématique: groupes et algèbres de Lie. Chapitre 9. Groupes de Lie réels compacts* (Masson, Paris, 1982)
8. T. Bröcker, T. tom Dieck, *Representations of Compact Lie Groups.* Graduate Texts in Mathematics, vol. 98 (Springer, New York, 1995)
9. B.M. Brown, Martingale central limit theorems. Ann. Math. Stat. **42**, 59–66 (1971)
10. M. Burger, *Horocycle flow on geometrically finite surfaces*, Duke Math. J. **61**(3), 779–803 (1990)
11. R. Burton, M. Denker, On the central limit theorem for dynamical systems. Trans. Am. Math. Soc. **302**, 715–726 (1987)
12. J.-P. Conze, Sur un critère de récurrence en dimension 2 pour les marches stationnaires, applications. Ergod. Theory Dyn. Syst. **19**, 1233–1245 (1999)
13. J.-P. Conze, S. Le Borgne, Méthode de martingales et flot géodésique sur une surface de courbure constante négative. Ergod. Theory Dyn. Syst. **21**(2), 421–441 (2001)
14. J.-P. Conze, S. Le Borgne, Limit directions of a vector cocycle, remarks and examples, Papers from the Probability and Ergodic Theory Workshops held at the University of North Carolina, Chapel Hill, NC, April 2014 (available on ArXiv)
15. F. Dal'Bo, *Trajectoires géodésiques et horocycliques* (EDP Sciences/CNRS Éditions, Paris, 2007)
16. D. Dolgopyat, On decay of correlations in Anosov flows. Ann. Math. (2) **147**(2), 357–390 (1998)
17. D. Dolgopyat, Limit theorems for partially hyperbolic systems. Trans. Am. Math. Soc. **356**(4), 1637–1689 (2004)
18. N. Enriquez, J. Franchi, Y. Le Jan, Central limit theorem for the geodesic flow associated with a Kleinian group, case $\delta > d/2$. J. Math. Pures Appl. (9), **80**(2), 153–175 (2001)
19. R. Fortet, Sur une suite également répartie. Stud. Math. **9**, 54–70 (1940). Polska Akademia Nauk. Instytut Matematyczny
20. J. Franchi, Y. Le Jan, *Hyperbolic Dynamics and Brownian Motion. An Introduction.* Oxford Mathematical Monographs (Oxford University Press, Oxford, 2012)
21. M. Gordin, The central limit theorem for stationary processes (English. Russian original). Sov. Math. Dokl. **10**(1969), 1174–1176 (1970). (Trans. from Dokl. Akad. Nauk SSSR **188**, 739–741 (1969))
22. S. Gouëzel, Central limit theorem and stable laws for intermittent maps. Probab. Theory Relat. Fields **128**(1), 82–122 (2004)
23. Y. Guivarc'h, Propriétés ergodiques, en mesure infinie, de certains systèmes fibrés. Ergod. Theory Dyn. Syst. **9**, 433–453 (1989)
24. Y. Guivarc'h, J. Hardy, Théorèmes limites pour une classe de chaînes de Markov et applications aux difféomorphismes d'Anosov. Ann. Inst. H. Poincaré Probab. Stat. **24**(1), 73–98 (1988)
25. P. Hall, C.C. Heyde, *Martingale Limit Theory and Its Application.* Probability and Mathematical Statistics (Academic, New York/London, 1980)
26. H. Hennion, L. Hervé, *Limit Theorems for Markov Chains and Stochastic Properties of Dynamical Systems by Quasi-compactness.* Lecture Notes in Mathematics, vol. 1766 (Springer, Berlin, 2001)
27. R. Howe, On a notion of rank for unitary representations of the classical groups, in *Harmonic Analysis and Group Representations* (Liguori, Naples, 1982), pp. 223–331
28. R. Howe, E.-E. Tan, *Nonabelian Harmonic Analysis. Applications of* SL(2, R). Universitext (Springer, New York, 1992)
29. C. Jan, Vitesse de convergence dans le TCL pour des processus associés à des systèmes dynamiques et aux produits de matrices aléatoires, Thèse, Université de Rennes 1, 2001
30. A. Katok, R.J. Spazier, First cohomology of Anosov actions of higher rank abelian groups and applications to rigidity. Publ. IHES **79**, 131–156 (1994)

31. Y. Katznelson, Ergodic automorphisms of $\mathbb{T}^n$ are Bernoulli shifts. Isr. J. Math. **10**, 186–195 (1971)
32. D.Y. Kleinbock, G.A. Margulis, Bounded orbits of nonquasiunipotent flows on homogeneous spaces, in *Sinai's Moscow Seminar on Dynamical Systems*. American Mathematical Society Translations Series 2, vol. 171 (American Mathematical Society, Providence, 1996), pp. 141–172
33. S. Le Borgne, Limit theorems for non-hyperbolic automorphisms of the torus. Isr. J. Math. **109**, 61–73 (1999)
34. S. Le Borgne, Principes d'invariance pour les flots diagonaux sur $SL(d, \mathbb{R})/SL(d, \mathbb{Z})$. Ann. Inst. H. Poincaré Probab. Stat. PR **38**(4), 581–612 (2002)
35. S. Le Borgne, F. Pène, Vitesse dans le théorème limite central pour certains systèmes dynamiques quasi-hyperboliques. Bull. Soc. Math. Fr. **133**(3), 395–417 (2005)
36. Y. Le Jan, The central limit theorem for the geodesic flow on noncompact manifolds of constant negative curvature. Duke Math. J. **74**(1), 159–175 (1994)
37. C. Liverani, Central limit theorem for deterministic systems, in *International Conference on Dynamical Systems*, Montevideo, 1995. Pitman Research Notes in Mathematics Series, vol. 362 (Longman, New York/Essex, 1996), pp. 56–75
38. C. Liverani, On contact Anosov flows. Ann. Math. (2) **159**, 1275–312 (2004)
39. I. Melbourne, A. Török, Central limit theorems and invariance principles for time-one maps of hyperbolic flows. Commun. Math. Phys. **229**(1), 57–71 (2002)
40. C.C. Moore, Exponential decay of correlation coefficients for geodesic flows, in *Group Representations, Ergodic Theory, Operator Algebras, and Mathematical Physics* (Berkeley, 1984). Mathematical Sciences Research Institute Publications, vol. 6 (Springer, New York, 1987), pp. 163–181
41. J.G. Ratcliffe, *Foundations of Hyperbolic Manifolds*. Graduate Texts in Mathematics, vol. 149 (Springer, New York, 2006)
42. M. Ratner, The central limit theorem for geodesic flows on n-dimensional manifolds of negative curvature. Isr. J. Math. **16**, 181–197 (1973)
43. M. Rees, Divergence type of some subgroups of finitely generated Fuchsian groups. Ergod. Theory Dyn. Syst. **1**(2), 209–221 (1981)
44. J. Rousseau-Egele, Un théoréme de la limite locale pour une classe de transformations dilatantes et monotones par morceaux. Ann. Probab. **11**(3), 772–788 (1983)
45. K. Schmidt, On joint recurrence. C. R. Acad. Sci. Paris, t. **327**, Série I, 837–842 (1998)
46. J.G. Sinaï, The central limit theorem for geodesic flows on manifolds of constant negative curvature. Dokl. Akad. Nauk SSSR **133**, 1303–1306 (1960) (Russian). (English translation in Soviet Math. Dokl. **1**, 983–987 (1960))
47. J.G. Sinaï, Markov partitions and C-diffeomorphisms. Funct. Anal. Appl. **2**, 61–82 (1968). (Trans. from Funkts. Anal. Prilozh. **2**(1), 64–89 (1968))
48. D. Volny, Counter examples to the central limit problem for stationary dependent random variables. Yokohama Math. J. **36**, 70–78 (1988)
49. L.-S. Young, Statistical properties of dynamical systems with some hyperbolicity. Ann. Math. **147**, 585–650 (1998)

Chapter 2
Semiclassical Approach for the Ruelle-Pollicott Spectrum of Hyperbolic Dynamics

Frédéric Faure and Masato Tsujii

Abstract Uniformly hyperbolic dynamics (Axiom A) have "sensitivity to initial conditions" and manifest "deterministic chaotic behavior", e.g. mixing, statistical properties etc. In the 1970, David Ruelle, Rufus Bowen and others have introduced a functional and spectral approach in order to study these dynamics which consists in describing the evolution not of individual trajectories but of functions, and observing the convergence towards equilibrium in the sense of distribution. This approach has progressed and these last years, it has been shown by V. Baladi, C. Liverani, M. Tsujii and others that this evolution operator ("transfer operator") has a discrete spectrum, called "Ruelle-Pollicott resonances" which describes the effective convergence and fluctuations towards equilibrium.

Due to hyperbolicity, the chaotic dynamics sends the information towards small scales (high Fourier modes) and technically it is convenient to use "semiclassical analysis" which permits to treat fast oscillating functions. More precisely it is appropriate to consider the dynamics lifted in the cotangent space T^*M of the initial manifold M (this is an Hamiltonian flow). We observe that at fixed energy, this lifted dynamics has a relatively compact non-wandering set called the trapped set and that this lifted dynamics on T^*M scatters on this trapped set. Then the existence and properties of the Ruelle-Pollicott spectrum enters in a more general theory of semiclassical analysis developed in the 1980 by B. Helffer and J. Sjöstrand called "quantum scattering on phase space".

We will present different models of hyperbolic dynamics and their Ruelle-Pollicott spectrum using this semi-classical approach, in particular the geodesic flow

F. Faure (✉)
Institut Fourier, UMR 5582, 100 rue des Maths, BP74 38402 St. Martin d'Hères, France
e-mail: frederic.faure@ujf-grenoble.fr

M. Tsujii
Department of Mathematics, Kyushu University, Moto-oka 744, Nishi-ku, Fukuoka, 819-0395, Japan
e-mail: tsujii@math.kyushu-u.ac.jp

F. Dal'Bo et al. (eds.), *Analytic and Probabilistic Approaches to Dynamics in Negative Curvature*, Springer INdAM Series 9, DOI 10.1007/978-3-319-04807-9_2,
© Springer International Publishing Switzerland 2014

on (non necessary constant) negative curvature surface $\mathcal{M}$. In that case the flow is on $M = T_1^* \mathcal{M}$, the unit cotangent bundle of $\mathcal{M}$. Using the trace formula of Atiyah-Bott, the spectrum is related to the set of periodic orbits.

We will also explain some recent results, that in the case of Contact Anosov flow, the Ruelle-Pollicott spectrum of the generator has a structure in vertical bands. This band spectrum gives an asymptotic expansion for dynamical correlation functions. Physically the interpretation is the emergence of a quantum dynamics from the classical fluctuations. This makes a connection with the field of quantum chaos and suggests many open questions.

2.1 Introduction

In these lecture notes, we present the use of semiclassical analysis for the study of hyperbolic dynamics. This approach is particularly useful in the case where the dynamics has neutral direction(s) like extensions of expanding maps, hyperbolic maps or Anosov flows.

In this approach we study the transfer operator associated to the dynamics and its spectral properties. The objective is to describe the discrete spectrum of the transfer operator, called "Ruelle-Pollicott resonances" and its importance to express the exponential time decay of correlation functions. This discrete spectrum (together with eigenvectors) is also useful to obtain further results for the dynamics as statistical results (central limit theorem, large deviations, linear response theory...), and to obtain estimates for counting of periodic orbits in the case of flow.

2.1.1 The General Idea Behind the Semiclassical Approach

1. Consider a smooth diffeomorphism $f : M \to M$ on a smooth manifold M (or a flow $f^t = \exp(tX) : M \to M, t \in \mathbb{R}$ generated by a vector field X). In the 1970, David Ruelle, Rufus Bowen and others have suggested to consider *evolution of functions* (resp. probability measures) with the pull back operator also called the transfer operator $\mathcal{L}^t \varphi = \varphi \circ f^{-t}$ (resp. its adjoint $\mathcal{L}^{t*}$) *instead of evolution of individual trajectories* $x(t) = f^t(x)$. This functional approach is useful for chaotic dynamical systems for which individual trajectories have unpredictable behavior, whereas a smooth density may converge towards equilibrium in a predictable manner.[1] Remark that this description is not reductive because taking $\varphi = \delta_x$ a Dirac measure at point x, one recovers the individual trajectory. See Fig. 2.1.

---

[1] Rem: this is somehow the weather is "predicted" by computer simulations from different initial conditions.

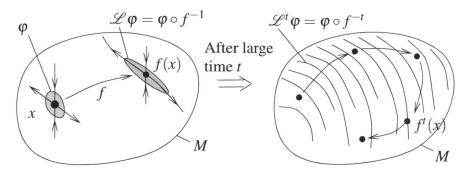

Fig. 2.1 An (hyperbolic) map f defines the evolution of a point $x \in M$ by $f^t(x)$ and evolution of a function $\varphi(x)$ by $\mathscr{L}^t \varphi = \varphi \circ f^{-t}$. The support of $\mathscr{L}^t \varphi$ spreads and folds after large time t

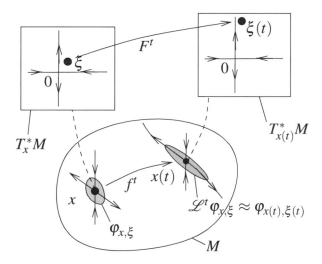

Fig. 2.2 Evolution of a wave packet

2. By linearity of the transfer operator $\mathscr{L}^t$, *a function (or distribution) on M can be decomposed as a superposition of "elementary wave packets"*[2] $\varphi_{x,\xi}$: this is a function with parameters (x, ξ) which has small support around $x \in M$ in space and whose Fourier transform (in local chart) also decay very fast outside some value $\xi \in T_x^* M$ in Fourier space.[3] Geometrically $(x, \xi) \in T^* M$ is a point on *the cotangent space*. A fundamental observation is that the time evolution of this wave packet $\mathscr{L}_t \varphi_{x,\xi}$ after finite time t, *remains a wave packet* with new parameters $(x(t), \xi(t)) = F^t(x, \xi) \in T^* M$ which follow the canonical lift $F : T^* M \to T^* M$ of the map $f : M \to M$. See Fig. 2.2.

---

[2]In signal theory and analysis this decomposition corresponds to wavelet transform or F.B.I. transform. In quantum physics an elementary wave packets is also called a "quantum".

[3]Fourier transform of φ is written $(\mathscr{F}\varphi)(\xi) = \frac{1}{(2\pi)^n} \int e^{-i\xi \cdot x} \varphi(x) \, dx$.

3. We therefore study the dynamics of the lift map $F^t : T^*M \to T^*M$. In the case of hyperbolic (Anosov) dynamics every point $(x(t), \xi(t))$ escape towards infinity $|\xi(t)| \to \infty$ as $t \to \pm\infty$, except if $(x(0), \xi(0)) \in K :=$ $\{(x, \xi), \xi = 0\}$, the zero section, called the "*trapped set*". A consequence is the *decay of correlation functions* $\left(\varphi_{x', \xi'}, \mathcal{L}_t \varphi_{x, \xi}\right)$ as $t \to \infty$ (intuitively only the constant function with $\xi = 0$ component survives). From the uncertainty principle in phase space T^*M this also implies that the transfer operator has *discrete spectrum* in some functional spaces "adapted" to the dynamics (so called Ruelle-Pollicott resonances). Here "adapted" means that the norm of this functional space has the ability to "truncate" the high frequencies. The limit of high frequencies $|\xi| \gg 1$ is called *the semiclassical limit*. Technically we will use semiclassical analysis and "quantum scattering theory" developed by Helffer-Sjöstrand and others in the 1980s [32] with "*escape functions*" (or Lyapounov function in phase space) in order to define these "*anisotropic Sobolev spaces*".

4. In the case of partially hyperbolic dynamics, e.g. Anosov vector field, then $|\xi(t)| \to \infty$ outside a "trapped set" $K \subset T^*M$ (or non wandering set) which is non compact. Geometrical properties of the trapped set K gives some more refined properties of the Ruelle-Pollicott spectrum of resonances, and also properties of the eigenspaces. For example its fractal dimension gives an (upper bound) estimate for the *density of Ruelle resonances*. If $K \subset T^*M$ is a symplectic submanifold this implies an *asymptotic spectral gap*, a *band structure* for the Ruelle spectrum, etc.

In order to present this approach we will consider different models. These models are very similar and the elaboration is increasing from one to the next. In particular we will present recent results for:

1. "*U(1) extension of Anosov diffeomorphism preserving a contact form*" [20, 23]. This model is also called *prequantum Anosov map*. It can be considered as a simplified model of a contact Anosov flow: there is a neutral direction for the dynamics and a contact one form that is preserved. This allows to obtain precise information on the Ruelle-Pollicott spectrum in the semiclassical limit of high frequencies along the neutral direction. In particular we will show that the spectrum has some *band structures* and obtain the "Weyl law" giving the number of resonances in each band. We will also show that surprisingly the correlation functions have some "*quantum behavior*". We will discuss the fact that these results propose a direct bridge between the study of Ruelle-Pollicott resonances in dynamics and questions in "quantum chaos" or "wave chaos". Using the Atiyah-Bott trace formula, we will relate the spectrum with the periodic orbits.

2. "*Contact Anosov flow*" [22, 24, 25]. This dynamical model can be considered as the analogous of the previous model in case of continuous time. This model is interesting in geometry because it includes the case of geodesic flow on a Riemannian manifold $\mathcal{M}$ with negative (sectional) curvature. In that case the flow takes place on the unit (co)tangent bundle $M = T_1^*\mathcal{M}$. We will show that all the results obtained for the previous model are also true here and concern

the spectrum of the generator of the flow (the vector field). We will discuss the relation with the spectrum of the Laplacian operator Δ on $\mathcal{M}$. We will express these results using zeta functions.

Sections or paragraphs marked with $\star$ can be skipped for a first lecture.

2.2 Hyperbolic Dynamics

2.2.1 Anosov Maps

Definition 2.1. On a C^∞ closed connected manifold M, a C^∞ diffeomorphism $f : M \to M$ is *Anosov* if there exists a Riemannian metric g on M, an f-invariant continuous decomposition of TM:

$$T_x M = E_u(x) \oplus E_s(x), \quad \forall x \in M, \tag{2.1}$$

a constant $\lambda > 1$ such that for every $x \in M$,

$$\forall v_s \in E_s(x), \quad \|D_x f(v_s)\|_g \leq \frac{1}{\lambda} \|v_s\|_g \tag{2.2}$$

$$\forall v_u \in E_u(x), \quad \left\|D_x f^{-1}(v_u)\right\|_g \leq \frac{1}{\lambda} \|v_u\|_g.$$

We call $E_u(x)$ the *unstable subspace* and $E_s(x)$ *the stable subspace*, see Fig. 2.3.

Example 2.1. Hyperbolic automorphism on the torus:

$$f : \begin{cases} \mathbb{T}^d := \mathbb{R}^d / \mathbb{Z}^d & \to \mathbb{T}^d \\ x & \to Mx \mod \mathbb{Z}^d, \end{cases} \tag{2.3}$$

with $M \in SL_d(\mathbb{Z})$ hyperbolic , i.e. every eigenvalues λ satisfy $|\lambda| \neq 1, 0$.

Remark 2.1.

- f in (2.3) is well defined because if $n \in \mathbb{Z}^d$, $x \in \mathbb{R}^d$ then

$$M(x+n) = Mx + \underbrace{Mn}_{\in \mathbb{Z}^d} = Mx \mod \mathbb{Z}^d.$$

- f is invertible on $\mathbb{T}^d$ and $f^{-1}(x) = M^{-1}x$ with $M^{-1} \in SL_d(\mathbb{Z})$.
- The simplest example of (2.3) is the "*cat map*" on $\mathbb{T}^2$[1] (cp. Fig. 2.4),

$$M = \begin{pmatrix} 2 & 1 \\ 1 & 1 \end{pmatrix}, \quad \lambda = \lambda_u = \frac{3 + \sqrt{5}}{2} \simeq 2.6 > 1, \quad \lambda_s = \lambda^{-1} < 1. \tag{2.4}$$

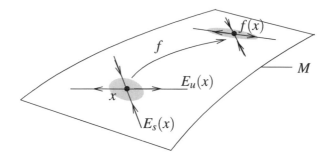

Fig. 2.3 An Anosov map f

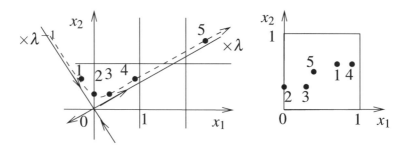

Fig. 2.4 Trajectory of an initial point $(-0.3, 0.6)$ under the cat map, on $\mathbb{R}^2$ (there the trajectory is on an hyperbola) and on $\mathbb{T}^2$. After restriction by modulo 1, the trajectory is "chaotic"

2.2.1.1 General Properties of Anosov Diffeomorphism

- In general, the maps $x \in M \rightarrow E_u(x), E_s(x)$ are not C^∞ but only Hölder continuous with some exponent $0 < \beta \leq 1$. (This is similar to the Weierstrass function).

★ It is conjectured that M is an infranil manifold. Ex: $M = \mathbb{T}^d$ is a torus.

Proposition 2.1 (★ "Structural stability" [34]). *If $f : M \rightarrow M$ is Anosov there exists $\varepsilon > 0$ such that for any $g : M \rightarrow M$ such that $\|g - \mathrm{Id}\|_{C^1} \leq \varepsilon$ then:*

1. $g \circ f$ is Anosov.
2. There exists an homeomorphism $h : M \rightarrow M$ (Hölder continuous) such that we have a commutative diagram:

$$
\begin{array}{ccc}
M & \xrightarrow{g \circ f} & M \\
\uparrow h & & \uparrow h \\
M & \xrightarrow{f} & M
\end{array}
$$

Proof. See [34]. The proof uses a description in terms on cones.

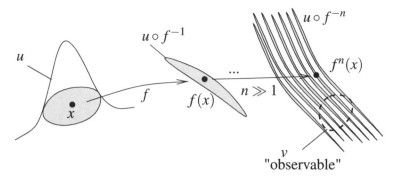

Fig. 2.5 The correlation function $C_{v,u}(n) := \int_M v.(u \circ f^{-n})\,dx$ represents the evolved function $u \circ f^{-n}$ tested against an "observable" function v

Theorem 2.1 (Anosov). *If* $f : M \to M$ *is Anosov and preserves a smooth measure* dx *on* M *then* f *is exponentially mixing:* $\exists \alpha > 0, \forall u, v \in C^\infty(M)$, *for* $n \to +\infty$,

$$\left| \underbrace{\int_M v.(u \circ f^{-n})\,dx}_{C_{v,u}(n)} - \int v\,dx.\int u\,dx \right| = O\left(e^{-\alpha n}\right) \tag{2.5}$$

In the last equation, the term

$$C_{v,u}(n) := \int_M v.(u \circ f^{-n})\,dx \tag{2.6}$$

is called a *correlation function.*

Remark 2.2. Mixing means "loss of information" because for $n \to \infty$, $u \circ f^{-n}$ normalized by $\left(\int u\,dx\right)^{-1}$ converges in the sense of distribution towards the measure dx. See Fig. 2.5.

Proof. This will be obtained in (2.47) as a consequence of Theorem 2.6, using semiclassical analysis. (From [26].) □

Remark 2.3. For *linear* Anosov map on $\mathbb{T}^d$, Eq. (2.3), the proof of exponentially mixing is easy and is true for any $\alpha > 0$. Let $k, l \in \mathbb{Z}^d$, let $\varphi_k(x) := \exp(i2\pi k.x)$ be a *Fourier mode*. Then

$$\int_{\mathbb{T}^d} (\varphi_k \circ f^{-n}).\overline{\varphi}_l\,dx = \int \exp(i2\pi(k.M^{-n}x - l.x))\,dx \tag{2.7}$$

$$= \int \exp\left(i2\pi\left({}^t M^{-n}k - l\right).x\right)dx = \delta_{{}^t M^{-n}k = l}.$$

But if $k \neq 0$ then $|^t M^{-n} k| \to \infty$ as $n \to +\infty$ because M is hyperbolic. So (2.7) vanishes for n large enough. Finally smooth functions u, v have Fourier components which decay fast and one deduces (2.5) for any $\alpha > 0$.

Proposition 2.2. ★ f *Anosov is ergodic:* $\forall u, v \in C^\infty (M)$,

$$\frac{1}{n} \sum_{k=0}^{n-1} \int v \cdot \left(u \circ f^{-k}\right) dx \xrightarrow[n \to \infty]{} \int v dx \cdot \int u dx. \qquad (2.8)$$

Proof. Using Cesaro's Theorem, one sees that mixing (2.5) implies ergodicity (2.8).

Remark 2.4. ★ Ergodicity means that the "time average" of v i.e. $\frac{1}{n} \sum_{k=0}^{n-1} \left(u \circ f^{-k}\right)$ normalized by $\left(\int u dx\right)^{-1}$ converges (in the sense of distribution) towards the measure dx.

Remark 2.5. Exponentially mixing (2.5) implies some *statistical properties* such as the central limit theorem for time average of functions etc.

2.2.2 Prequantum Anosov Maps

We introduce now "prequantum Anosov map": it is a $U(1)$ extension of an Anosov diffeomorphism f preserving a contact form. This corresponds to the "*geometric prequantization*" following Souriau-Kostant-Kirillov, Zelditch [46].

We will suppose that (M, ω) is a *symplectic manifold* and $f : M \to M$ is an Anosov map preserving ω:

$$f^* \omega = \omega \qquad (2.9)$$

i.e. f is symplectic. Then $\dim M = 2d$ is even and f preserves the non degenerate volume form $dx = \omega^{\wedge d}$ of degree $2d$.

Example 2.2. As (2.3) but with $f \in Sp_{2d}(\mathbb{Z}) : \mathbb{T}^{2d} \to \mathbb{T}^{2d}$ symplectic and hyperbolic. The linear cat map (2.4) is symplectic for $\omega = dq \wedge dp$ with coordinates $(q, p) \in \mathbb{R}^2$.

Remark 2.6. For every $x \in M$, $(T_x M, \omega)$ is a symplectic linear space (by definition) and $E_u(x), E_s(x) \subset T_x M$ given by (2.1) are *Lagrangian* linear subspaces hence

$$\dim E_u(x) = \dim E_s(x) = d.$$

Proof. If $u_s, v_s \in E_s(x)$ then

$$\omega(u_s, v_s) \underset{(2.9)}{=} \omega\left(D_x f^n(u_s), D_x f^n(v_s)\right) \xrightarrow[n \to \infty, (2.2)]{} 0.$$

Similarly for $E_u(x)$ with $D_x f^{-n}$.

Assumption 1:   The cohomology class $[\omega] \in H^2(M, \mathbb{R})$ represented by the symplectic form ω is integral, that is, $[\omega] \in H^2(M, \mathbb{Z})$.

Assumption 2:

(a) $H_1(M, \mathbb{Z}) \hookrightarrow H_1(M, \mathbb{R})$ is injective (i.e. no torsion part), and

(b) 1 is not an eigenvalue of the linear map $f_* : H_1(M, \mathbb{R}) \to H_1(M, \mathbb{R})$ induced by $f : M \to M$.

Remark 2.7. Assumption 1 is true for the cat map as $\int_{\mathbb{T}^2} \omega = \int_{\mathbb{T}^2} dq \wedge dp = 1 \in \mathbb{Z}$. Assumption 2-(b) is conjectured to be true for every Anosov map.

Theorem 2.2 ([23]). *With Assumption 1, there exists a $U(1)$-principal bundle $\pi : P \to M$ with connection one form $A \in C^\infty(P; \Lambda^1 \otimes i\mathbb{R})$ with curvature $\Theta = dA = -i(2\pi)(\pi^*\omega)$.*

With Assumption 2, we can choose the connection A above such that there exists a map $\tilde{f} : P \to P$ called prequantum map *(see Fig. 2.6) such that:*

1. The following diagram commutes:

$$
\begin{array}{ccc}
P & \xrightarrow{\tilde{f}} & P \\
\downarrow \pi & & \downarrow \pi \\
M & \xrightarrow{f} & M
\end{array}
\tag{2.10}
$$

2. "Equivariance" with respect to the action of $e^{i\theta} \in U(1)$:

$$
\forall p \in P, \forall \theta \in \mathbb{R}, \quad \tilde{f}\left(e^{i\theta} p\right) = e^{i\theta} \tilde{f}(p).
\tag{2.11}
$$

3. $\tilde{f}$ preserves the connection

$$
\tilde{f}^* A = A.
\tag{2.12}
$$

Proof. See [23].

Remark 2.8. At every point $p \in P$, $(\mathrm{Ker}A)(p) = \tilde{E}_u(p) \oplus \tilde{E}_s(p)$ is the strong distribution of stable/unstable directions of the map $\tilde{f}$. We recall the interpretation of the curvature two form Θ as an infinitesimal holonomy [42, (6.22), p. 506]. The fact that ω is symplectic here means that the distribution $\tilde{E}_u \oplus \tilde{E}_s$ is maximally "non integrable". The form $\alpha = \frac{i}{2\pi} A$ is a *contact one form* on P preserved by $\tilde{f}$ because

$$
\mu_P = \frac{1}{d!} \alpha \wedge (d\alpha)^d = \frac{1}{d!} \left(\frac{1}{2\pi} d\theta\right) \wedge \omega^d
$$

is a non degenerate $(2d+1)$ volume form on P preserved by $\tilde{f}$.

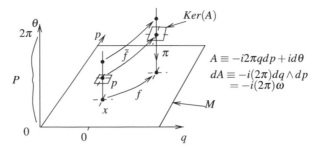

Fig. 2.6 A picture of the prequantum bundle $P \to M$ in the case of $M = \mathbb{T}^2$, e.g. for the "cat map" (2.4), with connection one form A and the prequantum map $\tilde{f} : P \to P$ which is a lift of $f : M \to M$. A fiber $P_x \equiv U(1)$ over $x \in M$ is represented here as a segment $\theta \in [0, 2\pi[$. The plane at a point p represents the horizontal space $H_p P = \mathrm{Ker}\left(A_p\right)$ which is preserved by $\tilde{f}$. These plane form a non integrable distribution with curvature given by the symplectic form ω

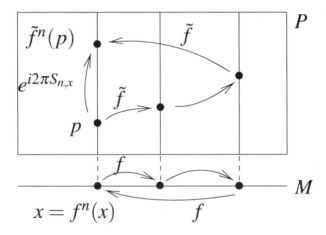

Fig. 2.7 Action of a periodic point $x = f^n(x)$

Remark 2.9. $\tilde{f}$ is a "partially hyperbolic map" with neutral direction θ, preserving a contact one form $\alpha = \frac{i}{2\pi} A$. Then $\tilde{f}$ is exponentially mixing (see Eq. (2.5)), but this is not obvious. This is a result of D. Dolgopyat [15]. We will obtain this in Remark 2.40 page 101.

Remark 2.10. ★ If $x = f^n(x)$ with $n \geq 1$, i.e. x is a periodic point of f, then for any $p \in P_x = \pi^{-1}(x)$,

$$\tilde{f}^n(p) = e^{i 2\pi S_{n,x}} p \tag{2.13}$$

with some phase $S_{n,x} \in \mathbb{R}/\mathbb{Z}$ called the *action* of the periodic point, cp. Fig. 2.7. This will appear in Trace formula in Sect. 2.4.

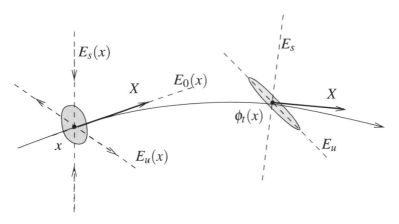

Fig. 2.8 Anosov flow

2.2.3 Anosov Vector Field

Definition 2.2. On a C^∞ manifold M, a smooth vector field X is *Anosov* (Fig. 2.8) if its flow $\phi_t = e^{-tX}$, $t \in \mathbb{R}$, satisfies:

1. $\forall x \in M$, we have an ϕ_t invariant decomposition:

$$T_x M = E_u(x) \oplus E_s(u) \oplus \underbrace{E_0(x)}_{\mathbb{R}X}. \qquad (2.14)$$

2. There exists a metric g on M, $\exists \gamma > 0$, $C > 0$, $\forall x \in M$, $\forall t \geq 0$,

$$\forall v_s \in E_s(x), \quad \|D_x\phi_t(v_s)\|_g \leq Ce^{-t\gamma} \|v_s\|_g \qquad (2.15)$$

$$\forall v_u \in E_u(x), \quad \|D_x\phi_{-t}(v_u)\|_g \leq Ce^{-t\gamma} \|v_u\|_g.$$

Remark 2.11. In general the maps $x \in M \to E_u(x)$, $E_s(x)$ are not smooth. They are only Hölder continuous with some exponent $0 < \beta \leq 1$.

Definition 2.3. We define the *Anosov one form* α on M by: $\forall x \in M$,

$$\alpha(E_u(x) \oplus E_s(x)) = 0, \qquad \alpha(X) = 1. \qquad (2.16)$$

In general $\alpha(x)$ is Hölder continuous with respect to $x \in M$. It is preserved by the flow: its Lie derivative is (in the sense of distributions) $\mathscr{L}_X \alpha = 0$. Conversely there is a unique one form α such that $\mathscr{L}_X \alpha = 0$ and $\alpha(X) = 1$.

Definition 2.4. $(\phi_t)_{t \in \mathbb{R}}$ is a *contact Anosov flow* if α is a C^∞ contact 1-form, i.e. if $\omega = d\alpha_{|E_u(x) \oplus E_s(x)}$ is nondegenerate (i.e. symplectic); equivalently $dx = \alpha \wedge (d\alpha)^d$ is an invariant smooth volume form on M, with $d = \dim E_u(x) = \dim E_s(x)$ (Lagrangian subspaces, see Remark 2.6). Then $\dim M = 2d + 1$.

Remark 2.12. That the flow is contact means that the distribution of hyperspaces $E_u(x) \oplus E_s(x)$ is maximally non integrable, this is similar to Fig. 2.6.

Example 2.3 ("geodesic flow with negative curvature").
  Let $\mathcal{M}$ be a smooth compact Riemannian manifold:

- The cotangent space $T^*\mathcal{M}$ has a canonical one form called the *Liouville one form* given by $\alpha = -\sum_{j=1}^n p^j dq^j$ in canonical coordinates (q^j are coordinates on $\mathcal{M}$ and p^j on $T_q^*\mathcal{M}$) [12]. The canonical symplectic form on $T^*\mathcal{M}$ is given by

$$\omega := \sum_j dq^j \wedge dp^j = d\alpha.$$

- On the cotangent space $T^*\mathcal{M}$, the Hamiltonian function $H(q,p) := \|p\|_g$ (with $p \in T_x\mathcal{M}$) defines a Hamiltonian vector field X by $\omega(X,.) = dH$ whose flow is called the *geodesic flow*. The energy level of energy 1 is the unit cotangent bundle $H^{-1}(1) = T_1^*\mathcal{M}$. The Hamiltonian flow preserves ω but also the one form α because $H(q,p)$ is homogeneous[4] in p. Therefore the geodesic flow is a contact flow on $M = T_1^*\mathcal{M}$ preserving α. The Anosov one form is $-\alpha$.

---

[4]*Proof.* Let

$$\mathcal{E} := \sum_j p^j \frac{\partial}{\partial p^j}$$

be the *canonical Euler vector field* on $T^*\mathcal{M}$ (it preserves fibers, it is canonically defined in any vector space). $\mathcal{E}$ generates the flow of "scaling":

$$S_\lambda : (q,p) \in T^*\mathcal{M} \to (q, e^\lambda p) \in T^*\mathcal{M}, \qquad \lambda \in \mathbb{R}. \qquad (2.17)$$

We have $\mathcal{E}(H) = H$ because H is homogeneous of degree 1 in p. We have $\iota_\mathcal{E}\alpha = 0$, $\iota_\mathcal{E}\omega = \alpha$ and $\mathscr{L}_\mathcal{E}\alpha = \alpha$, $\mathscr{L}_\mathcal{E}\omega = \omega$. The Hamiltonian vector field X_N is the associated *Reeb vector field*, i.e. it is uniquely defined by

$$\alpha(X) = -H = -1, \quad (d\alpha)(X) = 0. \qquad (2.18)$$

In particular X preserves α, i.e. $\mathscr{L}_X\alpha = 0$, i.e. it is a *contact vector field*. Indeed: we have on $T_E^*\mathcal{M}$

$$\alpha(X) = \iota_X\alpha = \iota_X(\iota_\mathcal{E}\omega) = -\omega(X,\mathcal{E}) = -\mathcal{E}(H) = -H = -E.$$

Also, $(d\alpha)(X) = \omega(X,.) = dH = 0$ on Σ_E. Then, on $T^*\mathcal{M}$,

$$\mathscr{L}_X\alpha = d(\iota_X\alpha) + \iota_X d\alpha = -d(H) + dH = 0. \qquad \square$$

- In the case where $\mathcal{M}$ has *negative sectional curvature* it is known that the geodesic flow is Anosov. This is therefore a *contact Anosov flow* on $M = T_1^* \mathcal{M}$. One has $\dim M = 2\dim \mathcal{M} - 1 = 2d + 1$. Therefore $n = \dim \mathcal{M} = d + 1$.

Example 2.4. A particular example is when $\mathcal{M}$ is a *homogeneous manifold*: $\mathcal{M} = \Gamma\backslash SO(1,n)/SO(n) = \Gamma\backslash\mathbb{H}^n$ where Γ is a discrete co-compact subgroup and $\mathbb{H}^n$ is the hyperbolic space of dimension n. The simplest case is when $\mathcal{M}$ is a surface ($n = \dim\mathcal{M} = 2$): one has $SO(2,1) \equiv SL_2(\mathbb{R})$. This case is explained in details below.

The following proposition shows how to obtain other contact (Anosov) vector field from a given one by "re-parametrization".

Proposition 2.3. ★ *If X_0 is a contact Anosov vector field with contact one form α_0, let β a closed one form on M such that $|\beta(X_0)| < \alpha_0(X_0) = 1$ then*

$$X = \frac{1}{1 + \beta(X_0)} X_0$$

is a also a contact Anosov vector field for the contact one form $\alpha = \alpha_0 + \beta$.

Proof. We have $d\alpha = d\alpha_0$ and

$$\alpha(X) = \frac{1}{1 + \beta(X_0)}(\alpha_0(X_0) + \beta(X_0)) = 1$$

and

$$\mathcal{L}_X\alpha = \iota_X d\alpha + d(\iota_X\alpha) = \frac{1}{1 + \beta(X_0)}\iota_{X_0}d\alpha_0 = \frac{1}{1 + \beta(X_0)}\mathcal{L}_{X_0}\alpha_0 = 0. \qquad \square$$

Remark 2.13. ★ P. Foulon and B. Hasselblatt [28] have shown that even in dimension 3 there are numerous contact Anosov flow that are not topologically orbit equivalent to geodesic flows.

Example 2.5 (Geodesic flow on a constant negative curvature surface.). We present here a standard example of contact Anosov flow, the geodesic flow on Riemann surface $\mathcal{M} = \Gamma\backslash(SL_2\mathbb{R}/SO_2)$ where $\Gamma < SL_2\mathbb{R}$ is a co-compact discrete subgroup. This example is a particular case of the Example 2.3 above. We present it in details, because we will use it later on in Sect. 2.3.5.1.

From Iwasawa decomposition, a matrix $g \in SL_2\mathbb{R}$ can be written[5]

$$g = \begin{pmatrix} y^{1/2} & x \\ 0 & y^{-1/2} \end{pmatrix} \begin{pmatrix} \cos\theta & \sin\theta \\ -\sin\theta & \cos\theta \end{pmatrix}, \quad x \in \mathbb{R}, y > 0, \theta \in SO_2.$$

---

[5]Recall that $g \in SL_2\mathbb{R} \Leftrightarrow \det g = 1$.

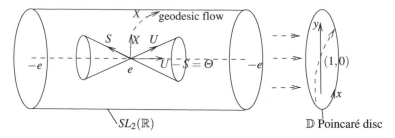

Fig. 2.9 Geodesic flow the Poincaré disc is generated by $X \in sl_2\mathbb{R}$

Hence with $z = x + iy \in \mathbb{H}^2$ in the Poincaré half plane, we have the homeomorphism $SL_2\mathbb{R} \equiv \mathbb{H}^2 \times SO_2$.

A basis of the Lie algebra $sl_2\mathbb{R} \equiv T_e (SL_2\mathbb{R})$ is[6]

$$X = \frac{1}{2}\begin{pmatrix} 1 & 0 \\ 0 & -1 \end{pmatrix}, \quad U = \begin{pmatrix} 0 & 1 \\ 0 & 0 \end{pmatrix}, \quad S = \begin{pmatrix} 0 & 0 \\ 1 & 0 \end{pmatrix}$$

and satisfies

$$[X, U] = U, \quad [X, S] = -S, \quad [U, S] = 2X. \tag{2.19}$$

These tangent vector X, U, S can be extended as left invariant vector fields on $SL_2\mathbb{R}$ by $X = g.X_e$ etc. Then the vector field X generates the flow $\phi_t = e^{-tX}$, cp. Fig. 2.9. It is given by the right action[7] of e^{-tX_e}: $\phi_t (g) := g.e^{-tX_e}$ and taking any left invariant metric g on $SL_2\mathbb{R}$ we have

$$\|D\phi_t (U)\|_g = \|U.e^{-tX}\|_g \underbrace{=}_{\|.\|_g l-inv.} \|e^{tX}.U.e^{-tX}\|_g = \|e^{t[X,.]}U\|_g \underset{(2.19)}{=} e^t \|U\|_g.$$
$$\tag{2.20}$$

According to (2.15), this shows that U spans the unstable direction $E_u (g)$ with $\lambda = e > 1$. Similarly we get $\|D\phi_t (S)\|_g = e^{-t} \|S\|_g$ and S spans $E_s (g)$. Therefore, if $\Gamma < SL_2\mathbb{R}$ is a discrete co-compact subgroup then $M := \Gamma\backslash SL_2\mathbb{R}$ is a compact manifold and X is a smooth contact Anosov vector field on M with $E_u = \mathbb{R}U$, $E_s = \mathbb{R}S$, $E_0 = \mathbb{R}X$. The property of contact comes from the last commutator $[U, S] = 2X$. More precisely the Anosov one form α, Eq. (2.16) is given by

$$\alpha = \frac{1}{2}K (X, .)$$

---

[6]Because $a \in sl_2\mathbb{R} \Leftrightarrow \det (e^a) = e^{\mathrm{Tr}a} = 1 \Leftrightarrow \mathrm{Tr}a = 0$.

[7]Indeed $\frac{d\phi_t}{dt}_{/t=0} = -g.X_e = -X$.

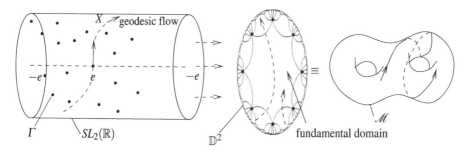

Fig. 2.10 Geodesic flow on a surface $\mathcal{M}$ with constant negative curvature

where $K = 2X^* \otimes X^* + 4\,(U^* \otimes S^* + S^* \otimes U^*)$ is the Killing metric on $SL_2\mathbb{R}$. To show this, observe that $\alpha\,(X) = 1$ and

$$(d\alpha)\,(S, U) = U\,(\alpha\,(S)) + S\,(\alpha\,(U)) - \alpha\,([S, U]) = -\frac{1}{2}K\,(X, [S, U])$$

$$= K\,(X, X) = 1$$

hence $d\alpha$ is symplectic on $E_u \oplus E_s = \mathrm{Span}\,(U, S)$.

If $(-\mathrm{Id}) \in \Gamma$ it is known that this flow can be identified with the geodesic flow on the Riemann surface $\mathcal{M} = \Gamma \setminus (SL_2\mathbb{R}/SO_2) = \Gamma \backslash \mathbb{H}^2$ which has constant negative curvature $\kappa = -1$ and that $M \equiv T_1^* \mathcal{M}$, see Fig. 2.10 below.

Remark 2.14. In $SL_2\mathbb{R} \equiv SO_{1,2}$ and similarly in higher dimension, some left invariant vector field on $M := \Gamma \backslash SO_{1,n}/SO_{n-1}$ are contact Anosov vector field and can be interpreted as the geodesic flow on a compact hyperbolic manifold $\mathcal{N} = \Gamma \backslash \mathbb{H}^n = M/SO_n$. The left invariant vector field on $\Gamma \backslash SO_{1,n}$ generates the frame flow.

2.2.3.1 ★ General Properties of Contact Anosov Flows

Theorem 2.3. *A contact Anosov flow is exponentially mixing, see Fig. 2.11: that is,* $\exists \alpha > 0$, $\forall u, v \in C^\infty\,(M)$, *for* $t \to \infty$ *one has*

$$\left| \underbrace{\int_M v.\,(u \circ \phi_{-t})\,dx - \int v dx.\int u dx}_{C_{v,u}(t)} \right| = O\left(e^{-\alpha t}\right). \tag{2.21}$$

The term $C_{v,u}\,(t)$ above is called a correlation function.

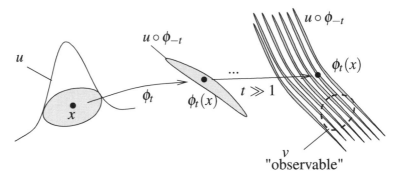

Fig. 2.11 Exponential mixing from the correlation function $C_{v,u}(t) = \int_M v. (u \circ \phi_{-t}) \, dx$

Remark 2.15. Mixing implies ergodicity. This is the same definition and same proof as in (2.8). Usually the term "correlation function" is for the whole difference $\int_M v. (u \circ \phi_{-t}) \, dx - \int v \, dx. \int u \, dx$.

2.3 Transfer Operators and Their Discrete Ruelle-Pollicott Spectrum

Before considering the Ruelle spectrum of Anosov dynamics, the following section introduces the techniques on a very simple example. This simple example (extended in $\mathbb{R}^d$) will also be important later on in the proof of Theorems 2.7 and 2.10 because it will serve as a universal "normal form".

2.3.1 Ruelle Spectrum for a Basic Model of Expanding Map

Let $\lambda > 1$ and consider the expanding map:

$$f : \begin{cases} \mathbb{R} & \to \mathbb{R} \\ x & \to \lambda x \end{cases}. \tag{2.22}$$

2.3.1.1 Transfer Operator

Let $u, v \in \mathscr{S}(\mathbb{R})$. The *time correlation function* (2.6) is for $n \geq 1$ (see Fig. 2.12):

$$C_{v,u}(n) := \int_{\mathbb{R}} \overline{v}. (u \circ f^{-n}) \, dx = \int \overline{v(x)}.u\left(\frac{x}{\lambda^n}\right) dx \xrightarrow[n \to +\infty]{} \left(\int \overline{v} dx\right).u(0). \tag{2.23}$$

Fig. 2.12 Illustration of the correlation function (2.23)

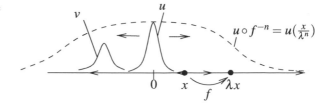

Let us write $\langle v|u\rangle_{L^2} := \int_{\mathbb{R}} \bar{v}.u\, dx$ for the L^2 scalar product. Let us define the *transfer operator*

$$\left(\hat{F}u\right)(x) := \left(u \circ f^{-1}\right)(x) = u\left(\frac{x}{\lambda}\right) \tag{2.24}$$

which is useful to express the correlation function:

$$C_{v,u}(n) = \int_{\mathbb{R}} \bar{v}.u \circ f^{-n} dx = \langle v|\hat{F}^n u\rangle_{L^2}.$$

Remark 2.16. The dual operator $\hat{F}^*$ defined by $\langle u|\hat{F}^*v\rangle = \langle \hat{F}u|v\rangle$ is given by[8]

$$\left(\hat{F}^*v\right)(y) = \lambda.v(\lambda y). \tag{2.25}$$

Taking $u = 1$ in $\langle u|\hat{F}^*v\rangle = \langle \hat{F}u|v\rangle$ gives that $\int \left(\hat{F}^*v\right)(x)\, dx = \int v(x)\, dx$. Hence $\hat{F}^*$ preserves probability measures. It is called the *Perron-Frobenius operator* or *Ruelle operator*.

2.3.1.2 Asymptotic Expansion

In this subsection we perform heuristic (non rigorous) computation in order to motivate the next section where these computations will be put in rigorous statements. The objective is to show the appearance and meaning of Ruelle spectrum of resonances. From Taylor formula (we don't care about the reminder for the moment) one has

$$u\left(\frac{x}{\lambda^n}\right) = \sum_{k\geq 0} \frac{x^k}{k!\lambda^{kn}} u^{(k)}(0).$$

---

[8]Proof: with the change of variable $y = \frac{x}{\lambda}$, we write

$$\langle u|\hat{F}^*v\rangle = \langle \hat{F}u|v\rangle = \int \overline{u\left(\frac{x}{\lambda}\right)}v(x)\, dx = \int \overline{u(y)}v(\lambda x)\,\lambda dy$$

hence $\left(\hat{F}^*v\right)(y) = \lambda.v(\lambda y)$.

Let $\delta^{(k)}$ be the k-th derivative of the Dirac distribution. Then

$$C_{v,u}(n) = \int \overline{v(x)}.u\left(\frac{x}{\lambda^n}\right)dx = \sum_{k \geq 0} \frac{1}{k!\lambda^{kn}}\left(\int x^k \overline{v(x)}dx\right).u^{(k)}(0)$$

$$= \sum_{k \geq 0} \frac{1}{\lambda^{kn}}\langle v|x^k\rangle\langle\frac{1}{k!}\delta^{(k)}|u\rangle$$

$$= \left(\int \overline{v}dx\right).u(0) + O\left(\frac{1}{\lambda^n}\right). \qquad (2.26)$$

We have[9] for $k, l \geq 0$

$$\langle\frac{1}{k!}\delta^{(k)}|x^l\rangle = \delta_{k=l}. \qquad (2.27)$$

Let[10]

$$\Pi_k := |x^k\rangle\langle\frac{1}{k!}\delta^{(k)}| \qquad (2.28)$$

be a rank one operator. Then (2.27) implies that

$$\Pi_k \circ \Pi_l = \delta_{k=l}.\Pi_k$$

i.e. $(\Pi_k)_k$ is a family of rank 1 projectors and the Taylor expansion (2.26) writes:

$$C_{v,u}(n) = \langle v|\hat{F}^n u\rangle = \sum_{k \geq 0} \frac{1}{(\lambda^k)^n}\langle v|\Pi_k u\rangle. \qquad (2.29)$$

Question 2.1. Formally this suggests the following spectral decomposition for the transfer operator $\hat{F}$:

$$``\hat{F} = \sum_{k \geq 0} \lambda^{-k}\Pi_k'', \qquad \hat{F}x^k = \lambda^{-k}x^k \qquad (2.30)$$

i.e. $\lambda_k = \lambda^{-k}$ should be "simple eigenvalues" and Π_k associated "spectral projector"; but in which space?

Notice that this statement cannot be true in the Hilbert space $L^2(\mathbb{R})$ because the distributions $x^k, \delta^{(k)}$ do not belong to it. The aim is to find an Hilbert space of

---

[9]Because $\left(\frac{d^k x^l}{dx^k}\right)(0) = 0$ if $k \neq l$ and $= k!$ if $k = l$.

[10]The expression $|x^k\rangle\langle\frac{1}{k!}\delta^{(k)}|$ is a notation (called "Dirac notation" in physics) for the rank one operator $x^k\langle\frac{1}{k!}\delta^{(k)}|.\rangle$.

distributions containing $\mathscr{S}(\mathbb{R})$ where the statement (2.30) holds true. We will have to consider Hilbert spaces as subspace of distributions. Notice first that the operator $\hat{F}$ defined in (2.24) can be extended by duality[11] to distributions $\hat{F} : \mathscr{S}'(\mathbb{R}) \rightarrow \mathscr{S}'(\mathbb{R})$.

Remark 2.17. The expanding map f in (2.22) is the time one flow $f = \phi_{t=1}$ generated by the vector field on $\mathbb{R}$

$$X = \gamma x \frac{d}{dx} \tag{2.32}$$

with $e^{\gamma} = \lambda > 1$. The transfer operator can be written in terms of the generator X:

$$\hat{F} = e^{-X}.$$

Remark 2.18. In $L^2(\mathbb{R})$ the operator $\frac{1}{\sqrt{\lambda}}\hat{F} = \frac{1}{\sqrt{\lambda}}e^{-X}$ is unitary and has continuous spectrum on the unit circle. Correspondingly the operator $i\left(X + \frac{\gamma}{2}\right)$ is selfadjoint in $L^2(\mathbb{R})$ and has continuous spectrum on $\mathbb{R}$. But as said above, we will not consider this Hilbert space.

2.3.1.3   Ruelle Spectrum

Theorem 2.4 ([23, Prop. 4.19]).   *For any $C > 0$, there exists a Hilbert space $\mathscr{H}_C$ (an "anisotropic Sobolev space" defined below)*

$$\mathscr{S}(\mathbb{R}) \subset \mathscr{H}_C \subset \mathscr{S}'(\mathbb{R})$$

such that the operator (2.24): $\hat{F} : \mathscr{H}_C \rightarrow \mathscr{H}_C$ is bounded and has essential spectral radius $r_{ess} = \text{cste} \cdot \lambda^{-C}$ ($\underset{C \rightarrow +\infty}{\rightarrow} 0$). The eigenvalues outside r_{ess} are $\lambda_k = \lambda^{-k}$ with $k \in \mathbb{N}$ and their spectral projector are $\Pi_k : \mathscr{H}_C \rightarrow \mathscr{H}_C$, given by Eq. (2.28). These eigenvalues $(\lambda_k)_{k \geq 0}$ are called Ruelle-Pollicott resonances. The generator $-X :$ $\mathscr{H}_C \rightarrow \mathscr{H}_C$ in (2.32) has discrete spectrum on $\text{Re}(z) > -C\gamma + \text{cste}$ ($\underset{C \rightarrow +\infty}{\rightarrow} -\infty$) and has eigenvalues $-k\gamma$, $k \in \mathbb{N}$. See Fig. 2.13.

A consequence is an expansion of correlation functions $C_{v,u}(n) = \langle v|\hat{F}^n u\rangle$ as (2.26) and (2.29) but with a controlled remainder:

---

[11]If $\alpha \in \mathscr{S}'(\mathbb{R})$, $\hat{F}\alpha$ is defined by

$$\forall u \in \mathscr{S}(\mathbb{R}), \quad \hat{F}(\alpha)(\bar{u}) = \langle u|\hat{F}\alpha\rangle = \langle \hat{F}^* u|\alpha\rangle = \alpha\left(\widehat{\hat{F}_v^*}(u)\right). \tag{2.31}$$

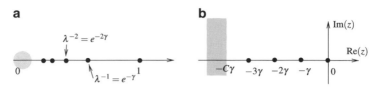

Fig. 2.13 (a) Spectrum of $\hat{F} = e^{-X} : \mathcal{H}_C \to \mathcal{H}_C$; (b) Spectrum of its generator $(-X) : \mathcal{H}_C \to \mathcal{H}_C$

Corollary 2.1. *For any $K \geq 0$, there exists $C_K > 0$, such that for any $u, v \in \mathcal{S}(\mathbb{R})$,*

$$\left| \langle v | \hat{F}^n u \rangle - \sum_{k=0}^{K} \frac{1}{(\lambda^k)^n} \langle v | \Pi_k u \rangle \right| \leq C_K \|v\|_{\mathcal{H}'_C} \|u\|_{\mathcal{H}_C} \frac{1}{(\lambda^{K+1})^n}.$$

Proof. ★ Let $K \geq 0$. Let $C \gg 0$ so that from Theorem 2.4 $r_{ess} < \frac{1}{(\lambda^{K+1})}$. Let $\hat{F} = \hat{K} + \hat{R}$ be a spectral decomposition in the space $\mathcal{H}_C$ with $r_{ess} < r_{spec}.(\hat{R}) < \frac{1}{(\lambda^{K+1})}$ and $\hat{K} = \sum_{k=0}^{K} \frac{1}{\lambda^k} \Pi_k$. Then $\hat{F}^n = \hat{K}^n + \hat{R}^n$ and

$$\langle v | \hat{F}^n u \rangle = \sum_{k=0}^{K} \frac{1}{(\lambda^k)^n} \langle v | \Pi_k u \rangle + \langle v | \hat{R}^n u \rangle.$$

We have $\left| \langle v | \hat{R}^n u \rangle \right| \leq \|v\|_{\mathcal{H}'_C} \|u\|_{\mathcal{H}_C} \left\| \hat{R}^n \right\|_{\mathcal{H}_C}$ and $\left\| \hat{R}^n \right\|_{\mathcal{H}_C}^{1/n} \xrightarrow[n\to\infty]{} r_{spec}.(\hat{R}) < \frac{1}{(\lambda^{K+1})}$ so $\left\| \hat{R}^n \right\|_{\mathcal{H}_C} \leq C_K \frac{1}{(\lambda^{K+1})^n}$. □

2.3.1.4 Arguments of Proof of Theorem 2.4

We will prove that $\hat{F} : \mathcal{H}_C \to \mathcal{H}_C$ has discrete spectrum. The proof presented below relies on a semiclassical approach, and is close to the proof of Theorem 1 in [26]. It also similar in spirit to the "quantum scattering theory in phase space" of by B. Helffer, J. Sjöstrand [32]. The same strategy will be used for Anosov maps in Sects. 2.3.2, 2.3.3 and Anosov flows in Sect. 2.3.4. The proof uses the "semiclassical theory of PDO" (cf. Appendix) and the idea behind is decomposition in wavepackets as explained in the introduction. The proof in [23] is closer to this idea. Before, let us give some important remarks.

Remark 2.19. The transfer operator is $\left(\hat{F}u\right)(x) = u\left(\frac{1}{\lambda}x\right)$. Let us consider the Fourier transform

Fig. 2.14 The canonical map
F, Eq. (2.33)

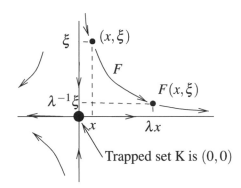

$$\tilde{u}(\xi) := \frac{1}{\sqrt{2\pi}} \int e^{-i\xi x} u(x)\, dx.$$

Then[12]

$$\widetilde{\left(\hat{F}u\right)}(\xi) = \lambda \tilde{u}(\lambda\xi).$$

Geometrically (x, ξ) are coordinates on the cotangent space $T^*\mathbb{R} \equiv \mathbb{R}^2$. This shows that u et $\tilde{u}$ are "transported" by the following canonical map $F : T^*\mathbb{R} \to T^*\mathbb{R}$ in the cotangent space $T^*\mathbb{R}$:

$$F : (x, \xi) \to \left(\lambda x, \lambda^{-1}\xi\right). \tag{2.33}$$

The map F is the canonical lift of the map $f : \mathbb{R} \to \mathbb{R}$.

We observe that the map F has a *trapped set* (or non wandering set) $K = (0, 0)$ *compact* in $T^*\mathbb{R}$ (cp. Fig. 2.14), in the precise sense that

$$K := \{(x, \xi), \exists C \Subset T^*M \text{ compact}, \forall n \in \mathbb{Z}, F^n(x, \xi) \in C\} = \{(0, 0)\}.$$

Remark 2.20. The dynamics of the map F in $\mathbb{R}^2 \equiv T^*\mathbb{R}$ looks like *"scattering"* on the trapped set K.

Remark 2.21. In the cotangent space $T^*\mathbb{R}$, the wave front (see Definition 2.16) of the distribution x^k which enters in the spectral projector (2.28) is the line

---

[12]Proof: $\widetilde{\left(\hat{F}u\right)}(\xi) = \frac{1}{\sqrt{2\pi}} \int e^{-i\xi x} u\left(\frac{1}{\lambda}x\right) dx = \lambda \frac{1}{\sqrt{2\pi}} \int e^{-i\xi\lambda y} u(y)\, dy = \lambda \tilde{u}(\lambda\xi).$

$$E_u = \{(x, \xi), x \in \mathbb{R}, \xi = 0\}$$

and the wavefront set of $\delta^{(k)}$ is the line

$$E_s = \{(x, \xi), x = 0, \xi \in \mathbb{R}\}.$$

They are respectively the unstable/stable manifolds for the trapped set K of the canonical map F.

Write $z := (x, \xi) \in \mathbb{R}^2$. For $C > 0$, consider the *Lyapounov function* or *escape function*, that is the C^∞ function

$$A_C(z) := \langle z \rangle^{m(z)} \tag{2.34}$$

where $\langle z \rangle := \sqrt{1 + |z|^2}$ and $m(z) \in C^\infty(\mathbb{R}^2)$ is the *order function*: a homogeneous function of degree 0 on $|z| \geq 1$ (that is $m(\lambda z) = m(z)$ for $|z| \geq 1$, $\lambda \geq 1$) such that $m(z) = +C$ in a conical vicinity of the stable axis $x = 0$, $m(z) = -C$ in a conical vicinity of the unstable axis $\xi = 0$, with $m(z)$ decreasing between these two directions so that

$$m(F(z)) \leq m(z), \forall |z| \geq 1. \tag{2.35}$$

Along the stable direction one has $|z| \sim |\xi| \gg 1$ and from (2.34) and (2.33) one has

$$A_C(z) \sim |\xi|^C, \qquad \frac{A_C(F(z))}{A_C(z)} \simeq \frac{|\lambda^{-1}\xi|^C}{|\xi|^C} \simeq \lambda^{-C} \ll 1.$$

Similarly along the unstable direction, one has $|z| \sim |x| \gg 1$ and

$$A_C(z) \sim |x|^{-C}, \qquad \frac{A_C(F(z))}{A_C(z)} \simeq \frac{|\lambda x|^{-C}}{|x|^{-C}} \simeq \lambda^{-C} \ll 1.$$

One can check in fact that in every direction and for $|z| \gg 1$ one has

$$\frac{A_C(F(z))}{A_C(z)} \lesssim \lambda^{-C} \ll 1. \tag{2.36}$$

Remark 2.22. ★ The function $m(z) \in S^0(\mathbb{R}^2)$ is a symbol according to (2.128) and the function $A_C \in S_\rho^{m(z)}$ is a symbol with variable order $m(z)$ according to (2.129), with any $0 < \rho < 1$.

Let us define the pseudodifferential operator (PDO) $\mathrm{Op}(A_C) : \mathscr{S}(\mathbb{R}) \to \mathscr{S}(\mathbb{R})$ by ordinary quantization (see Appendix 2.5)

$$(\text{Op}(A_C) u)(x) := \frac{1}{2\pi} \int e^{i\xi x} A_C (x, \xi) e^{-i\xi y} u (y) d\xi dy$$

(it can be modified by a subleading PDO, i.e. with lower order, so that it becomes selfadjoint and invertible). Then in $L^2 (\mathbb{R})$, let us consider the operator obtained by conjugation:

$$\hat{Q} := \text{Op}(A_C) \circ \hat{F} \circ \text{Op}(A_C)^{-1}.$$

From *Egorov Theorem* we have that

$$\text{Op}(A_C) \circ \hat{F} \circ \text{Op}(A_C)^{-1} = \hat{F} \circ \left(\text{Op}(A_C \circ F) + O \left(\text{Op}\left(S^{m \circ F - \rho} \right) \right) \right) \circ \text{Op}(A_C)^{-1}$$

where $(A_C \circ F) \in S^{m \circ F}$, the notation $O \left(\text{Op}\left(S^{m'} \right) \right)$ means a term which belongs to $\text{Op}\left(S^{m'} \right)$ and for any $1/2 < \rho < 1$. The *Theorem of composition of PDO* (see Appendix 2.5) gives that

$$\text{Op}(A_C \circ F) \circ \text{Op}(A_C)^{-1} = \text{Op}\left(\frac{A_C \circ F}{A_C} \right) + O \left(\text{Op}\left(S^{m \circ F - m - \rho} \right) \right)$$

where

$$\frac{A_C \circ F}{A_C} \in S^{m \circ F - m} \subset S^0.$$

The last inclusion is because $m \circ F - m \leq 0$ from (2.35).

In conclusion we have that

$$\hat{Q} = \hat{F} \circ \left(\text{Op}\left(\frac{A_C \circ F}{A_C} \right) + O \left(\text{Op}(S^{-\rho}) \right) \right).$$

The *theorem of L^2-continuity* gives that for norm operator

$$\left\| \text{Op}\left(\frac{A_C \circ F}{A_C} \right) + O \left(\text{Op}(S^{-\infty}) \right) \right\| \leq \limsup_{(x,\xi)} \left| \frac{A_C \circ F}{A_C} (x, \xi) \right| \underset{(2.36)}{\leq} \lambda^{-C}.$$

Since $\hat{F}$ is bounded on $L^2 (\mathbb{R})$ we have that $\left\| \hat{Q} + O \left(\text{Op}(S^{-\rho}) \right) \right\| \leq \left\| \hat{F} \right\|_{L^2(\mathbb{R})} \lambda^{-C}$. Finally an operator $\hat{K} \in \text{Op}(S^{-\rho})$ with $\rho > 0$ is compact hence

$$\hat{Q} = \hat{K} + \hat{R}$$

with $\left\| \hat{R} \right\| \leq \text{cst} \times \lambda^{-C}$ and $\hat{K}$ a *compact operator*. From the commutative diagram

$$L^2(\mathbb{R}) \xrightarrow{\hat{Q}} L^2(\mathbb{R})$$

$$\text{Op}(A_C) \uparrow \qquad\qquad \uparrow \text{Op}(A_C) \qquad\qquad (2.37)$$

$$\mathscr{H}_C \xrightarrow{\hat{F}} \mathscr{H}_C$$

one has the same result for $\hat{F}$ in the space

$$\mathscr{H}_C := \text{Op}(A_C)^{-1}\left(L^2(\mathbb{R})\right)$$

with norm $\|u\|_{\mathscr{H}_C} := \|\text{Op}(A_C)u\|_{L^2}$. The space $\mathscr{H}_C$ is called[13] *anisotropic Sobolev space*. Notice that $\mathscr{H}_C$ contains regular (smooth) functions but that may grows in x. So $x^k \in \mathscr{H}_C$ for $k \le C$, but $\delta^{(k)} \notin \mathscr{H}_C$. For the dual space $\mathscr{H}'_C = \text{Op}(A_C)\left(L^2(\mathbb{R})\right) = \mathscr{H}_{-C}$ this is the opposite: $\delta^{(k)} \in \mathscr{H}_{-C}$. As a result, the operator Π_k is bounded in $\mathscr{H}_C \to \mathscr{H}_C$.

Remark 2.23. ⋆ The dual operator (2.25) (or Perron Frobenius operator)

$$\hat{F}^* : \begin{cases} \mathscr{H}_{-C} & \to \mathscr{H}_{-C} \\ v & \to \lambda.v(\lambda x) \end{cases}$$

has the same spectrum λ^{-k}, $k \ge 0$. (conjugate spectrum, but the spectrum is real).

Remark 2.24. In a finite dimensional vector space a conjugation like (2.37) does not change the spectrum of the operator. In our case, with infinite dimension, the essential spectrum is moved away, and reveals discrete (Ruelle) spectrum that is "robust and intrinsic".

2.3.1.5 Ruelle Spectrum for Expanding Map in $\mathbb{R}^d$

Theorem 2.4 can be easily generalized for an expanding linear map on $\mathbb{R}^d$ with any $d \ge 1$. We will use this later.

Let $A : \mathbb{R}^d \to \mathbb{R}^d$ be a linear invertible expanding map satisfying $\|A^{-1}\| \le 1/\lambda$ for some $\lambda > 1$. Let

$$\mathscr{L}_A : \begin{cases} \mathscr{S}(\mathbb{R}^d) & \to \mathscr{S}(\mathbb{R}^d) \\ u & \to u \circ A^{-1} \end{cases} \qquad (2.38)$$

be the associated transfer operator. For $k \in \mathbb{N}$, let[14]

$$\text{Polynom}^{(k)} := \text{Span}\left\{x^\alpha, \alpha \in \mathbb{N}^d, |\alpha| = k\right\}$$

---

[13]Recall that the usual Sobolev space with constant order $m \in \mathbb{R}$ is defined by [41] as $H^m(\mathbb{R}) := \left(\text{Op}\left(\langle\xi\rangle^m\right)\right)^{-1}\left(L^2(\mathbb{R})\right)$.

[14]For a multi-index $\alpha \in \mathbb{N}^d$, $\alpha = (\alpha_1, \ldots \alpha_d)$, we write $|\alpha| = \alpha_1 + \ldots + \alpha_d$.

be the space of homogeneous polynomial on $\mathbb{R}^d$ of degree k.

$$\dim\left(\text{Polynom}^{(k)}\right) = \binom{d+k-1}{d-1} = \frac{(d+k-1)!}{(d-1)!k!}.$$

Then we consider the finite rank operator

$$\Pi_k : \mathscr{S}\left(\mathbb{R}^d\right) \to \text{Polynom}^{(k)}, \qquad (\Pi_k u)(x) = \sum_{\alpha \in \mathbb{N}^d, |\alpha|=k} \frac{\partial^\alpha u(0)}{\alpha!} x^\alpha. \qquad (2.39)$$

This is a projector which extracts the terms of degree k in the Taylor expansion.

We have the following relations

$$\Pi_j \circ \Pi_k = \delta_{j=k} \Pi_k \qquad (2.40)$$

and

$$[\Pi_k, \mathscr{L}_A] = 0. \qquad (2.41)$$

Let us prepare some notations. For a linear invertible map L we will use the notation

$$\|L\|_{\max} := \|L\|, \qquad \|L\|_{\min} := \left\|L^{-1}\right\|^{-1}. \qquad (2.42)$$

Theorem 2.5 ([23, Prop. 4.19]). *For any $C > 0$, there exists a Hilbert space $\mathscr{H}_C$ (an "anisotropic Sobolev space")*

$$\mathscr{S}\left(\mathbb{R}^d\right) \subset \mathscr{H}_C \subset \mathscr{S}'\left(\mathbb{R}^d\right)$$

such that the operator (2.38): $\mathscr{L}_A : \mathscr{H}_C \to \mathscr{H}_C$ is bounded and has essential spectral radius $r_{ess} = \text{cste}.\lambda^{-C} \left(\underset{C\to+\infty}{\to} 0\right)$. For $K \leq C - 2d$, there is a decomposition preserved by $\mathscr{L}_A$:

$$\mathscr{H}_C = \left(\bigoplus_{k=0}^{K} \text{Polynom}^{(k)}\right) \oplus \tilde{\mathscr{H}}$$

such that

1. $\exists C_0$, for any $0 \leq k \leq K$ and $0 \neq u \in \text{Polynom}^{(k)}$, we have for any $n \geq 1$,

$$C_0^{-1} \|A^n\|_{\max}^{-k} \leq \frac{\|\mathscr{L}_{A^n} u\|_{\mathscr{H}_C}}{\|u\|_{\mathscr{H}_C}} \leq C_0 \|A^n\|_{\min}^{-k}. \qquad (2.43)$$

2. The operator norm of the restriction of $\mathscr{L}_A$ to $\tilde{\mathscr{H}}$ is bounded by

$$C_0 \max\{\|A^n\|_{\min}^{-(K+1)}, \|A^n\|_{\min}^{-C} \cdot |\det A^n|\}. \qquad (2.44)$$

Remark 2.25. ★ Theorem 2.5 implies that the spectrum of the transfer operator $\mathscr{L}_A$ in the Hilbert space $\mathscr{H}_C$ is discrete outside the radius r_{ess}. The eigenvalues outside this radius are given by the action of $\mathscr{L}_A$ in the finite dimensional space Polynom$^{(k)}$. These eigenvalues can be computed explicitly from the Jordan block decomposition of A. In particular if $A = \mathrm{Diag}\,(a_1, \ldots a_d)$ is diagonal then the monomials $x^\alpha = x_1^{\alpha_1} \ldots x_d^{\alpha_d}$ are obviously eigenvectors of $\mathscr{L}_A$ with respective eigenvalues $\prod_j a_j^{-\alpha_j}$.

2.3.2 Ruelle Spectrum of Anosov map

Let $f : M \to M$ be an Anosov map as in Definition (2.1).

Definition 2.5. Let $V \in C^\infty (M)$ real valued, called potential. The transfer operator is

$$\hat{F} : \begin{cases} C^\infty (M) & \to C^\infty (M) \\ u & \to e^V \cdot \left(u \circ f^{-1}\right) \end{cases}. \tag{2.45}$$

Remark 2.26. The choice $u \circ f^{-1}$ instead of $u \circ f$ is such that f maps supp (u) to supp $\left(\hat{F}u\right)$.

Remark 2.27. ★The L^2 adjoint operator is given by $\left(\hat{F}^*v\right)(y) = \frac{e^{V \circ f}}{|\det Df|}(v \circ f)$ and called *Perron-Frobenius operator*. It transports densities and preserves probabilities if $V = 0$:

$$\int_M \left(\hat{F}^*v\right) dy = \langle 1|\hat{F}^*v\rangle = \langle \hat{F}1|v\rangle = \int v dy.$$

Remark 2.28. By duality the transfer operator can be extended to distributions: $\hat{F} : \mathscr{D}'(M) \to \mathscr{D}'(M)$.
Let $T^*M = E_s^* \oplus E_u^*$ be the decomposition dual to Eq. (2.1), i.e. $E_s^*(E_s) = 0$ and $E_u^*(E_u) = 0$.

Theorem 2.6 ([7, 9, 26, 39] *Discrete spectrum*). *For any $C > 0$, there exists an anisotropic Sobolev space $\mathscr{H}_C$:*

$$C^\infty (M) \subset \mathscr{H}_C \subset \mathscr{D}'(M)$$

with variable order function $m \in C^\infty (T^*M)$ with $m(x, \xi) = \pm C$ along $E_{u/s}^*$ such that

$$\hat{F} : \mathscr{H}_C \to \mathscr{H}_C$$

is bounded and has essential spectral radius $r_{ess} = O(1).\lambda^{-C}$ ($\underset{C \to +\infty}{\longrightarrow} 0$).

Fig. 2.15 Ruelle Pollicott resonances of $\hat{F}$

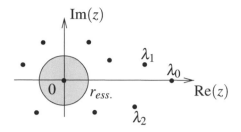

The eigenvalues (and eigenspaces) outside r_{ess} do not depend on m and are called Ruelle-Pollicott resonances (Fig. 2.15). The space $\mathcal{H}_C$ does not depend on V. The wavefront set of the eigendistributions is contained in E_u^*.

Remark 2.29. We will denote $\mathrm{Res}\left(\hat{F}\right)$ the set of Ruelle-Pollicott resonances (eigenvalues). The only obvious eigenvalue is for the case $V = 0$: it is $\lambda_0 = 1$ with eigenfunction $u_0 = 1$.

Remark 2.30. For the hyperbolic automorphism on the torus (2.3), with $V = 0$, the Ruelle spectrum is only $\mathrm{Res}\left(\hat{F}\right) = \{1\}$. To show this, use (2.7) and (2.46).

Remark 2.31. The Ruelle spectrum describes asymptotic of time correlation functions (2.6): for $V = 0$ in (2.45), one has for $u, v \in C^\infty(M)$ and any $\varepsilon > 0$,

$$C_{v,u}(n) \underset{(2.6)}{=} \int v.u \circ f^{-n} dx \underset{(2.45)}{=} \langle v | \hat{F}^n u \rangle$$

$$= \sum_{\lambda_j \in \mathrm{Res}(\hat{f}), |\lambda_j| \geq \varepsilon} \langle v | \left(\hat{F}^n \Pi_j \right) u \rangle + \|u\|_{\mathcal{H}_C} \cdot \|v\|_{\mathcal{H}_{-C}} \cdot O\left(\varepsilon^n\right) \quad (2.46)$$

where Π_j denotes the finite rank spectral projector $\hat{F}$ associated to the eigenvalue λ_j. $\mathcal{H}_{-C}$ is the space dual to $\mathcal{H}_C$ (precisely defined with the order function $-m(x, \xi)$ instead of $+m(x, \xi)$).

Proposition 2.4 (Anosov). *If $f : M \rightarrow M$ is an Anosov diffeomorphism preserving a smooth measure dx, then for any real valued potential V, there is a simple "leading" eigenvalue $\lambda_0 > 0$ in the sense that the other ones are $\lambda_j \in \mathbb{C}$ with $|\lambda_j| < \lambda_0$ as in Fig. 2.15.*

Remark 2.32. In the particular case $V = 0$ then $\lambda_0 = 1$, $\Pi_0 = |1\rangle\langle 1|$ and $|\lambda_1| < 1$. Then (2.46) gives that for any $\varepsilon > |\lambda_1|$:

$$C_{v,u}(n) = \int v.u \circ f^{-n} dx = \langle v | 1 \rangle \langle 1 | u \rangle + O\left(\varepsilon^n\right)$$

$$= \int \bar{v} dx. \int u dx + O\left(\varepsilon^n\right). \quad (2.47)$$

This proves the exponential mixing (2.5).

2.3.2.1 Proof of Theorem 2.6

This proof from [26] uses Semiclassical analysis. The proof is very similar to the proof of Theorem 2.4 given above.

The transfer operator (2.45) is a Fourier integral operator. Its canonical map is

$$F : \begin{cases} T^*M & \to T^*M \\ (x, \xi) & \to (x', \xi') = \left(f(x), {}^t Df_x^{-1}.\xi \right) \end{cases}. \tag{2.48}$$

F is the canonical lift of $f : M \to M$ on the cotangent bundle T^*M.

Heuristic Interpretation of the Canonical Map F from the Expression of the Transfer Operator $\hat{F}$ (2.45)

- If $u \in C^\infty(M)$ with support supp(u) then $\hat{F}u$ as support f (supp(u)). This explains that $x' = f(x)$ in (2.48).
- If on some local chart $u(x) = e^{i\xi.x}$ with some $|\xi| \gg 1$, i.e. u is a "fast oscillating function", then $\left(\hat{F}u \right)(y) = e^V e^{i\xi.f^{-1}(y)}$. Put $y = f(x) + y'$ with $|y'| \ll 1$, so $f^{-1}(y) = x + Df_y^{-1}.y' + o(|y'|)$ (by Taylor) so

$$\left(\hat{F}u \right)(y) \simeq e^V e^{i\xi.(x+Df_y^{-1}.y)} = C.e^{i\left({}^t Df^{-1}\xi \right).y} = C.e^{i\xi'.y}$$

with $\xi' = {}^t Df^{-1}\xi$. We have obtained (2.48).

The *trapped set* (or non wandering set) of the map $F : T^*M \to T^*M$ is the zero section

$$K = \{(x, \xi) \in T^*M, x \in M, \xi = 0\}. \tag{2.49}$$

For $\rho \in K$ we let (Fig. 2.16)

$$E_u^*(\rho) := \left\{ v \in T_\rho(T^*M), \left| DF_\rho^{-n}(v) \right| \xrightarrow[n \to +\infty]{} 0 \right\}$$

$$E_s^*(\rho) := \left\{ v \in T_\rho(T^*M), \left| DF_\rho^{n}(v) \right| \xrightarrow[n \to +\infty]{} 0 \right\}.$$

We define an *escape function* with *variable order* $m \in C^\infty(T^*M)$ so that

$$\left(\frac{A_m \circ F}{A_m} \right)(x, \xi) < C.\lambda^{-C} \ll 1 \quad \text{for } |\xi| \gg 1, \tag{2.50}$$

and such that $A_m \in S_\rho^m$ is a "good symbol" (see Definition 2.11). For this we choose

Fig. 2.16 The canonical map
F, Eq. (2.48)

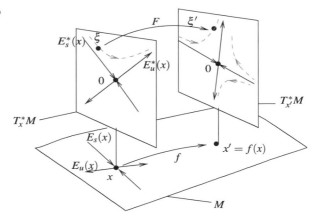

$$A_m (x, \xi) := \langle \xi \rangle^{m(x,\xi)} \tag{2.51}$$

with

$$m (x, \xi) = C \gg 0 \quad \text{along } E_s^*$$
$$m (x, \xi) = -C \ll 0 \quad \text{along } E_u^*.$$

Define the following *pseudo-differential operator* using local coordinates

$$\hat{A}_m u := \text{Op} (A_m) u := \frac{1}{(2\pi)^{2d}} \int e^{i\xi.x} A_m (x, \xi) e^{-i\xi.y} u (y) \, dy d\xi$$

and the *anisotropic Sobolev space:*

$$\mathcal{H}_C := \hat{A}_m^{-1} (L^2 (M)).$$

Then one has a commutative diagram:

$$
\begin{array}{ccc}
L^2 (\mathbb{R}) & \xrightarrow{\hat{Q} := \hat{A}_m \hat{F} \hat{A}_m^{-1}} & L^2 (\mathbb{R}) \\
\hat{A}_m \uparrow & & \hat{A}_m \uparrow \\
\mathcal{H}_C & \xrightarrow{\hat{F}} & \mathcal{H}_C
\end{array}
$$

Then:

$$\hat{Q} := \text{Op} (A_m) \circ \hat{F} \circ \text{Op} (A_m)^{-1} = \hat{F} \circ \text{Op} (A_m \circ F) \circ \text{Op} (A_m)^{-1} + l.o.t.$$

$$= \hat{F} \circ \text{Op} \left(\frac{A_m \circ F}{A_m} \right) + l.o.t.$$

From L^2 continuity theorem and (2.50), on has

$$\mathrm{Op}\left(\frac{A_m \circ F}{A_m}\right) = \hat{K} + \hat{R}$$

with $\left\|\hat{R}\right\| \leq c.\lambda^{-C}$ and $\hat{K}$ a *compact operator* (smoothing). The same decomposition holds for $\hat{Q} : L^2 \to L^2$ and $\hat{F} : \mathscr{H}_C \to \mathscr{H}_C$.

2.3.2.2 The Atiyah-Bott Trace Formula

Definition 2.6. The *flat trace* of the transfer operator (2.45) is

$$\mathrm{Tr}^\flat \hat{F} := \int_M K\left(x, x\right) dx \tag{2.52}$$

where $K\left(x, y\right) dy$ is the Schwartz kernel of $\hat{F}$.

Remark 2.33. We recall that the Schwartz kernel of $\hat{F}$ is defined by $\left(\hat{F}u\right)(x) = \int K\left(x, y\right) u\left(y\right) dy$. It is a current. More generally the flat trace can be defined for a vector bundle map $B : E \to E$ lifting a diffeomorphism $f : M \to M$ on a vector bundle $E \to M$, such that all fixed points of f are hyperbolic.

Proposition 2.5 ([4]). *For any $n \geq 1$, the* Atiyah-Bott trace formula *is*

$$\mathrm{Tr}^\flat\left(\hat{F}^n\right) = \sum_{x = f^n(x)} \frac{e^{V_n(x)}}{\left|\det\left(1 - Df_x^{-n}\right)\right|} \tag{2.53}$$

where

$$V_n\left(x\right) = \sum_{k=0}^{n-1} V\left(f^{-k}\left(x\right)\right). \tag{2.54}$$

Remark 2.34. In (2.53), this is a finite sum over periodic points.

Proof (* Atiyah-Bott [3, 4]). From (2.45), and denoting $\delta_y\left(x\right) = \delta\left(y - x\right)$ the Dirac distribution at y, the Schwartz kernel of $\hat{F}^n$ is

$$K_n\left(x, y\right) = \left(\hat{F}\delta_y\right)(x) = \delta_y\left(f^{-n}\left(x\right)\right) e^{V_n(x)}$$

$$= \delta\left(y - f^{-n}\left(x\right)\right) e^{V_n(x)}.$$

From (2.52), one has (using the change of variable $y = x - f^{-n}\left(x\right)$ in the second line)

$$\mathrm{Tr}^{\flat}\left(\hat{F}^{n}\right) = \int_{M} \delta\left(x - f^{-n}(x)\right) e^{V_{n}(x)} dx = \sum_{x=f^{n}(x)} \frac{e^{V_{n}(x)}}{\left|\det\left(1 - Df_{x}^{-n}\right)\right|}. \qquad \square$$

Remark 2.35. If f preserves dx then $\left|\det\left(Df_{x}^{n}\right)\right| = 1$ so $\mathrm{Tr}^{\flat}\left(\hat{F}^{n}\right) = \sum_{x=f^{n}(x)} \frac{e^{V_{n}(x)}}{\left|\det\left(1 - Df_{x}^{n}\right)\right|}$.

Lemma 2.1 ([6] Flat trace and spectrum). *For any $\varepsilon > 0$, we have*

$$\mathrm{Tr}^{\flat}\left(\hat{F}^{n}\right) = \sum_{\lambda_{j} \in \mathrm{Res}(\hat{F}),|\lambda_{j}| \geq \varepsilon} \lambda_{j}^{n} + O(1).\varepsilon^{n} \tag{2.55}$$

$$\underset{(2.53)}{=} \sum_{x=f^{n}(x)} \frac{e^{V_{n}(x)}}{\left|\det\left(1 - Df_{x}^{-n}\right)\right|}. \tag{2.56}$$

Proof. From [6] (see also [23, chap.11]) we decompose

$$\hat{F}^{n} = \hat{F}_{0}^{n} + \hat{F}_{1}^{n}$$

where $\hat{F}_{0} = \sum_{\lambda_{j},|\lambda_{j}| \geq \varepsilon} \hat{F}_{0} \Pi_{j}$ is the finite rank spectral component of $\hat{F}$, $\hat{F}_{1} = \hat{F} - \hat{F}_{0}$ so $\left[\hat{F}_{0}, \hat{F}_{1}\right] = 0$. One has $\left\|\hat{F}_{1}^{n}\right\| \leq O(1).\varepsilon^{n}$ and prove that

$$\left|\mathrm{Tr}^{\flat}\left(\hat{F}_{1}^{n}\right)\right| \leq O(1) \times \left\|\hat{F}_{1}^{n}\right\| \leq O(1)\varepsilon^{n}. \qquad \square$$

Consequences

As in Proposition 2.4 let $\lambda_{0} > 0$ be the leading eigenvalue and $|\lambda_{1}| < \lambda_{0}$ the next one. One has for any $\varepsilon > 0$,

$$\mathrm{Tr}^{\flat}\left(\hat{F}^{n}\right) \underset{(2.55)}{=} \lambda_{0}^{n} + O(1)(|\lambda_{1}| + \varepsilon)^{n}$$

$$\underset{(2.53)}{=} \sum_{x=f^{n}(x)} \frac{e^{V_{n}(x)}}{\left|\det\left(1 - Df_{x}^{-n}\right)\right|}$$

so for $n \gg 1$,

$$\log \lambda_{0} = \frac{1}{n} \log\left(\sum_{x=f^{n}(x)} \frac{e^{V_{n}(x)}}{\left|\det\left(1 - Df_{x}^{-n}\right)\right|}\right) + O(1)\left(\frac{(|\lambda_{1}| + \varepsilon)}{\lambda_{0}}\right)^{n}.$$

For a function $\varphi \in C(M)$ let

$$\Pr(\varphi) := \lim_{n \to \infty} \frac{1}{n} \log \left(\sum_{x = f^n(x)} e^{\varphi_n(x)} \right) \tag{2.57}$$

called *the topological pressure* of φ with $\varphi_n := \sum_{k=0}^{n-1} \varphi \left(f^{-k}(x) \right)$.

Using other transfer operators and because

$$\left| \det \left(1 - Df_x^{-n} \right) \right|^{-1} \underset{n \to \infty}{\sim} \left| \det Df_{|E_s(x)}^{-n} \right|^{-1} = e^{-J_n(x)}$$

with "*the unstable Jacobian*"[15]

$$J(x) := \log \left| \det Df_{|E_s(x)}^{-1} \right| \tag{2.58}$$

on can show that

Proposition 2.6. *One has:*

$$\log \lambda_0 = \Pr(V - J) \tag{2.59}$$

- *In particular in the case $V = 0$, we have $\lambda_0 = 1$ from Remark 2.32, so (2.59) gives $\Pr(-J) = 0$.*
- *In the particular case $V = J$, $\lambda_0 = \Pr(0) =: h_{top}$ is called the* topological entropy. *From Eq. (2.57), h_{top} gives the exponential rate for the number of periodic points:*

$$\sharp \{ x = f^n(x) \} \underset{n \to \infty}{\sim} e^{(h_{top} + o(1))n}.$$

2.3.3 Ruelle Band Spectrum for Prequantum Anosov Maps

Consider the prequantum map $\tilde{f} : P \to P$ defined in (2.10). We follow Sect. 2.3.2.

Definition 2.7. Let $V \in C^\infty(M)$ real valued, called *potential*. The prequantum transfer operator is

$$\hat{F} : \begin{cases} C^\infty(P) & \to C^\infty(P) \\ u & \to e^{V \circ \pi} \cdot \left(u \circ \tilde{f}^{-1} \right) \end{cases} . \tag{2.60}$$

It preserves the Fourier N-mode space for every $N \in \mathbb{Z}$:

---

[15]Notice that if $\det Df = 1$ then $\det Df_{|E_s(x)}^{-1} = \det Df_{|E_u(x)}$.

$$C_N^\infty (P) := \left\{ u \in C^\infty (P), \forall p \in P, \forall e^{i\theta} \in U(1), \quad u\left(e^{i\theta} p\right) = e^{iN\theta} u(p) \right\}.$$
(2.61)

We denote

$$\hat{F}_N := \hat{F}_{|C_N^\infty(P)}$$
(2.62)

and for $N \neq 0$ we put

$$\hbar := \frac{1}{2\pi N}.$$
(2.63)

Remark 2.36. From (2.63) we have $e^{iN\theta} = e^{i\theta/(2\pi\hbar)}$. This notation is used in quantum mechanics. So the space $C_N^\infty (P)$ contains functions which oscillates fast along the direction $\frac{\partial}{\partial\theta}$ as $N \to \infty$. For that reason, the limit $N \to \infty$ or $\hbar \to 0$ is called the *semiclassical limit*.

★ In the theory of associated vector bundles it is shown that [42]

$$C_N^\infty (P) \equiv C^\infty \left(M, L^{\otimes N}\right)$$

the space of sections of a complex line bundle power N, where $L \to M$ is the associated complex line bundle usually called *"prequantum line bundle"*. For simplicity it is equivalent to work with (2.61).

Remark 2.37. Theorem 2.6 extends to transfer operators acting on vector bundles. So it applies for the operator $\hat{F}_N$, for any N. Consequently the operator $\hat{F}_N$ has discrete spectrum of Ruelle resonances $\text{Res}\left(\hat{F}_N\right)$.

We define the special "potential of reference"

$$V_0(x) := \frac{1}{2} \log \left| \det Df_{f^{-1}(x)}|_{E_u(f^{-1}(x))} \right|.$$
(2.64)

Notice that the unstable foliation $E_u(x)$ is not smooth in x in general which implies that the function V_0 is Hölder continuous but not smooth in x. We then consider the difference

$$D := V - V_0 \quad \in C^\beta (M)$$
(2.65)

which is also a Hölder continuous function on M and that will be called the *"effective damping function"*. It will appear in many results below. Finally we denote by

$$D_n(x) := \sum_{j=1}^{n} D\left(f^j(x)\right)$$
(2.66)

the Birkhoff sum of the damping function. Recall (2.42) for the definition of $\|\cdot\|_{\max}$ and $\|\cdot\|_{\min}$.

Fig. 2.17 Band structure for
the Ruelle-Pollicott
resonances of $\hat{F}_N$

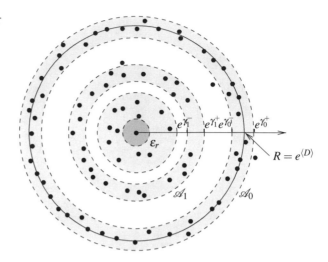

Theorem 2.7 ([23] Band Structure, Fig. 2.17). *For any $\varepsilon > 0$, there exists $C_\varepsilon > 0$, $N_\varepsilon \geq 1$ such that for any $N \geq N_\varepsilon$:*

1. *The Ruelle-Pollicott resonances of $\hat{F}_N$ are contained in a small neighborhood of the union of annuli $\left(\mathscr{A}_k := \{ r_k^- \leq |z| \leq r_k^+ \} \right)_{k \geq 0}$:*

$$Res\left(\hat{F}_N\right) \subset \bigcup_{k \geq 0} \underbrace{\{ r_k^- - \varepsilon \leq |z| \leq r_k^+ + \varepsilon \}}_{\varepsilon\text{-neighborhood of } \mathscr{A}_k} \tag{2.67}$$

   with

$$r_k^- := \liminf_{n \to \infty} \inf_{x \in M} \left(e^{\frac{1}{n} D_n(x)} \| Df_x^n |_{E_u} \|_{\max}^{-k/n} \right), \tag{2.68}$$

$$r_k^+ := \limsup_{n \to \infty} \sup_{x \in M} \left(e^{\frac{1}{n} D_n(x)} \| Df_x^n |_{E_u} \|_{\min}^{-k/n} \right).$$

2. *Suppose that $r_k^+ < r_{k-1}^-$ for some $k \geq 1$. For any $z \in \mathbb{C}$ such that*

$$r_k^+ + \varepsilon < |z| < r_{k-1}^- - \varepsilon$$

   i.e. such that z is in a "gap", the resolvent of $\hat{F}_N$ on $\mathscr{H}_N^r(P)$ is controlled uniformly with respect to N:

$$\left\| \left(z - \hat{F}_N \right)^{-1} \right\| \leq C_\varepsilon \tag{2.69}$$

   This is also true for $|z| > r_0^+ + \varepsilon$.
3. *If $r_1^+ < r_0^-$, i.e. if the outmost annulus $\mathscr{A}_0$ is isolated from other annuli, then the number of resonances in its neighborhood satisfies the estimate called "Weyl formula"*

Fig. 2.18 With the particular potential
$V_0 = \frac{1}{2} \log \left| \det Df_x|_{E_u(x)} \right|$
the external spectrum of the transfer operator $\hat{F}_N$ concentrates uniformly on the unit circle as
$N = 1/(2\pi\hbar) \to \infty$ (We have not represented here the structure of the internal bands inside the disc of radius r_1^+)

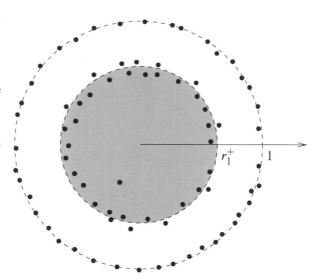

$$\sharp \left\{ Res\left(\hat{F}_N\right) \bigcap \left\{ r_0^- - \varepsilon \le |z| \le r_0^+ + \varepsilon \right\} \right\} = N^d \mathrm{Vol}_\omega(M)\left(1 + O\left(N^{-1}\right)\right)$$
(2.70)

with $\mathrm{Vol}_\omega(M) := \int_M \frac{1}{d!}\omega^{\wedge d}$ *being the symplectic volume of M and* $\delta > 0$. *Moreover in the limit* $N \to \infty$, *most of these resonances concentrate and equidistribute on the circle of radius*

$$R := e^{\langle D \rangle}, \quad with \ \langle D \rangle := \frac{1}{\mathrm{Vol}_\omega(M)} \int_M D(x)\,dx. \tag{2.71}$$

Remark 2.38. 1. Since $\|Df_x^n|_{E_u}\|_{\max}^{1/n} \ge \|Df_x^n|_{E_u}\|_{\min}^{1/n} > \lambda > 1$, from (2.2), we have obviously $r_k^- \le r_k^+$, $r_{k+1}^- < r_k^-$ and $r_{k+1}^+ < r_k^+$ for every $k \ge 0$. However we don't always have $r_{k+1}^+ < r_k^-$ therefore the annuli $\mathscr{A}_k$ may intersect each other.

2. In the case $V = 0$, one has $r_0^+ < 1$ so one can deduce exponential mixing for the prequantum map $\tilde{f}$, see Remark 2.9.

3. It is tempting to take the potential $V = V_0$ defined in (2.64) which would indeed give $D = 0$ hence $r_0^+ = r_0^- = 1$ in (2.68). In that case the external band $\mathscr{A}_0$ would be the unit circle, separated from the internal band $\mathscr{A}_1$ by a spectral gap r_1^+ given by

$$r_1^+ = \limsup_{n \to \infty} \sup_{x \in M} \left(\|Df_x^n|_{E_u}\|_{\min}^{-1/n} \right) < \frac{1}{\lambda} < 1$$

See Fig. 2.18. However Theorem 2.7 does not apply in this case because the function V_0 is not smooth in x as required. In [23] it is shown how to generalize the result to this case using an extension of the transfer operator to the Grassmanian bundle.

4. In the simple case of a linear hyperbolic map on the torus $\mathbb{T}^2$, i.e. Eq. (2.4) with $V(x) = 0$, then $r_k^+ = r_k^- = \lambda^{-k-\frac{1}{2}}$, with $\lambda = Df_{0/E_u} = \frac{3+\sqrt{5}}{2} \simeq 2.6$ (constant), i.e. each annulus $\mathscr{A}_k$ is a circle. In this case Theorem 2.7 has been obtained in [20, fig.1-b]. If one chooses $V(x) = \frac{1}{2} \log |\det Df_x|_{E_u}| = \frac{1}{2} \log \lambda$ the external band $\mathscr{A}_0$ is the unit circle and it is shown in [20] that the Ruelle-Pollicott resonances on the external band coincide with the spectrum of the quantized map called the "quantum cat map".

5. There is a conjecture of Pollicott and Dolgopyat [16] for a better estimate of r_0^+ in (2.67) in terms of the pressure (2.57) and J in (2.58):

$$\log r_0^+ = \frac{1}{2} \Pr(2V - 2J).$$

Definition 2.8. Suppose $r_1^+ < r_0^-$ (isolated external band). Let $\varepsilon > 0$, and $N_\varepsilon \geq 1$ given by Theorem 2.7. Let $\Pi_\hbar$ be the spectral projector on the external band $\mathscr{A}_0$ which is finite rank from (2.70). Let

$$\mathscr{H}_\hbar := \mathrm{Im}\,(\Pi_\hbar) \tag{2.72}$$

that we call the *"quantum space"* which is finite dimensional and let

$$\hat{\mathscr{F}}_\hbar : \mathscr{H}_\hbar \to \mathscr{H}_\hbar \tag{2.73}$$

be the finite dimensional spectral restriction of $\hat{F}_N$. We call $\hat{\mathscr{F}}_\hbar$ the *"quantum operator"*.

In fact, for every N we define Π_N as the spectral projector $|z| > r_1^+ + \varepsilon$ and put $\hat{\mathscr{F}}_N := \hat{F}_N \Pi_N$. In particular for $N \geq N_\varepsilon$ $\Pi_N = \Pi_\hbar$ and $\hat{\mathscr{F}}_N = \hat{\mathscr{F}}_\hbar$.

Theorem 2.8 (Correlation functions and interpretation [23]). *With the same setting as in the previous definition, for any $u, v \in C^\infty(P)$, and for $n \to \infty$, one has*

$$\underbrace{\left(v, \hat{F}^n u\right)_{L^2}}_{\text{"classical"}} = \sum_N \underbrace{\left(v_N, \hat{\mathscr{F}}_N^n u_N\right)}_{\text{"quantum"}} + O\left(\left(r_1^+ + \varepsilon\right)^n\right) \tag{2.74}$$

where $u_N, v_N \in C_N^\infty(P)$ are the Fourier components of the functions u and v. In the right hand side of (2.74), the sum is infinite but convergent.

Remark 2.39. Equation (2.74) has a nice interpretation: the classical correlation functions $\left(v, \hat{F}^n u\right)$ are governed by the quantum correlation functions $\left(v_N, \hat{\mathscr{F}}_N^n u_N\right)$ for large time, or equivalently the *"quantum dynamics emerge dynamically from the classical dynamics"*.

Remark 2.40. It is known that for $n \to \infty$,

$$\left(v, \hat{F}^n u\right) = \lambda_0^n \left(v, \Pi_{\lambda_0} u\right) + O\left(|\lambda_1|^n\right)$$

where $\lambda_0 > 0$ is the leading and simple eigenvalue of $\tilde{F}$ (in the space $\mathcal{H}_{N=0}^r$) and λ_1 is the second eigenvalue with $|\lambda_1| < \lambda_0$. The case $V = 0$ for which $\lambda_0 = 1$ gives that the map $\tilde{f} : P \to P$ is mixing with exponential decay of correlations.

Remark 2.41. ★ In [23] we show that $\hat{\mathcal{F}}_\hbar$ is a valuable quantization of the symplectic map f but different from usual "*geometric quantization*".

2.3.3.1 Proof of Theorem 2.7

The idea is the same as in the proof in Sect. 2.3.2.1 page 92, but we use now $\hbar$-semiclassical analysis with $\hbar := 1/(2\pi N) \ll 1$.

We consider charts $U_\alpha \subset M$ and local trivializations of the bundle P:

$$\tau_\alpha : U_\alpha \subset M \to P$$

i.e. diffeomorphisms

$$T_\alpha : \begin{cases} U_\alpha \times \mathbf{U}(1) & \to \pi^{-1}(U_\alpha) \\ (x, e^{i\theta}) & \to e^{i\theta} \tau_\alpha(x) \end{cases}. \tag{2.75}$$

Consequently the pull-back of the connection A on P by the trivialization map (2.75) is written as

$$T_\alpha^* A = id\theta - i2\pi \eta_\alpha \tag{2.76}$$

where $\eta_\alpha \in C^\infty\left(U_\alpha, \Lambda^1\right)$ is a one-form on U_α which depends on the choice of the local section τ_α. We have

$$\omega = d\eta_\alpha. \tag{2.77}$$

Lemma 2.2 (Local expression of the prequantum map $\tilde{f}$, Fig. 2.19). *Suppose that $V \subset U_\alpha \cap f^{-1}(U_\beta)$ is a simply connected open set. We have*

$$\tilde{f}(\tau_\alpha(x)) = e^{i2\pi \mathcal{A}_{\beta,\alpha}(x)} \tau_\beta(f(x)) \tag{2.78}$$

with the "action function" given by

$$\mathcal{A}_{\beta,\alpha}(x) = \int_{f(\gamma)} \eta_\beta - \int_\gamma \eta_\alpha + c(x_0) = \int_\gamma \left(f^*(\eta_\beta) - \eta_\alpha\right) + c(x_0). \tag{2.79}$$

Fig. 2.19 Illustrates the expression (2.78) of the prequantum map $\tilde{f}$ with respect to local trivialization. It is characterized by the action function $\mathscr{A}_{\beta,\alpha}(x)$

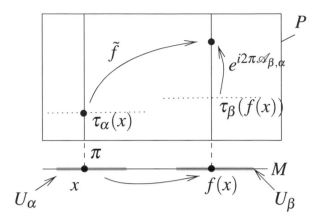

In the last integral, $x_0 \in V$ is any point of reference, $\gamma \subset V$ is a path from x_0 to x and $c(x_0)$ does not depend on x. See Fig. 2.19.

Lemma 2.3 (Local expression of $\hat{F}_N$). Let $u \in C_N^\infty(P)$ and $u' := \hat{F}_N u \in C_N^\infty(P)$. Let the respective associated functions be $u_\alpha = u \circ \tau_\alpha$ and $u'_\alpha = u' \circ \tau_\alpha$ for any indices α. Then

$$u'_\beta = e^V \cdot e^{-i2\pi N \mathscr{A}_{\beta,\alpha} \circ f^{-1}} \left(u_\alpha \circ f^{-1} \right). \tag{2.80}$$

Proposition 2.7 ($\hat{F}_N$ is a $\hbar$-Fourier Integral Operator). *Its local canonical map is*

$$F_{\alpha,\beta} : \begin{cases} T^*U_\alpha & \to T^*U_\beta \\ (x,\xi) & \to (x',\xi') = \left(f(x), {}^t\left(Df_{x'}^{-1}\right)(\xi + \eta_\alpha(x)) - \eta_\beta(x') \right) \end{cases} \tag{2.81}$$

where $x \in U_\alpha$, $f(x) \in U_\beta$ and $\xi \in T_x^*U_\alpha$. The map $F_{\alpha,\beta}$ preserves the canonical symplectic structure

$$\Omega := \sum_{j=1}^{2d} dx_j \wedge d\xi_j. \tag{2.82}$$

Proof. This comes from (2.80). See explanation of (2.48). There is a new term in (2.80): the multiplication operator by a "fast oscillating phase" (recall that $\hbar \ll 1$):

$$\hat{F}_2: \quad u(x) \to u'(x) = e^{iS(x)/\hbar} u(x)$$

with $S(x) = -\mathscr{A}_{\beta\alpha} \circ f^{-1} = \int_{f^{-1}(\gamma)} \eta_\alpha - \int_\gamma \eta_\beta - c(x_0)$. If $u(x) = e^{\frac{i}{\hbar}\xi \cdot x}$ then it is transformed to

$$u'(y) = \left(\hat{F}_2 u\right)(y) = e^{\frac{i}{\hbar}(\xi \cdot y + S(y))}$$

and for $y = x + y'$ with $|y'| \ll 1$, we have

$$u'(y) \simeq C e^{\frac{i}{\hbar}(\xi \cdot y + dS \cdot y)} = C e^{\frac{i}{\hbar}\xi' \cdot y}$$

with $\xi' = \xi + dS$, $C = e^{\frac{i}{\hbar}(S(x) - dS_x \cdot x)}$ and $dS = f^{-1*}\eta_\alpha - \eta_\beta$. This gives (2.81). □

Lemma 2.4. *With the following change of variable*

$$(x, \xi) \in T^* U_\alpha \to (x, \zeta) = (x, \xi + \eta_\alpha(x)) \in T^* M, \qquad (2.83)$$

the canonical map (2.81) get the simpler and global expression

$$F : \begin{cases} T^* M & \to T^* M \\ (x, \zeta) & \to (x', \zeta') = \left(f(x), {}^t\left(Df_{x'}^{-1}\right)\zeta\right) \end{cases} \qquad (2.84)$$

similar to (2.48), but the symplectic form Ω in (2.82) preserved by F is:

$$\Omega = \sum_{j=1}^{2d} \left(dx_j \wedge d\zeta_j\right) + \tilde{\pi}^*(\omega) \qquad (2.85)$$

with the canonical projection map $\tilde{\pi} : T^* M \to M$.

So as in (2.49), the trapped set is the zero section

$$K = \{(x, \xi) \in T^* M, x \in M, \xi = 0\} \subset T^* M. \qquad (2.86)$$

Here $(K, \Omega) \equiv (M, \omega)$ is a symplectic submanifold.
For every $\rho \in K$, we can decompose Ω orthogonally:

$$T_\rho(T^* M) = T_\rho K \overset{\perp_\Omega}{\bigoplus} (T_\rho K)^{\perp_\Omega}. \qquad (2.87)$$

Moreover

$$T_\rho K = \underbrace{E_u^{(1)} \oplus E_s^{(1)}}_{2d}, \qquad (T_\rho K)^\perp = \underbrace{E_u^{(2)} \oplus E_s^{(2)}}_{2d}$$

with $E_u^{(1)} := T_\rho K \cap E_u^*(\rho)$, etc.

With respect to the decomposition (2.87), the canonical map F is within the linear approximation

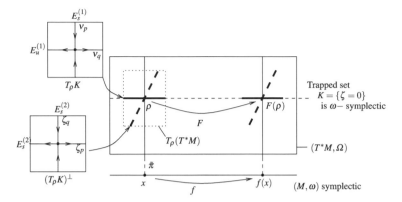

Fig. 2.20 The decompositions of the tangent space $T_\rho(T^*M)$

$$T_\rho\left(T^*M\right) = \underbrace{E_u^{(1)}(\rho) \oplus E_s^{(1)}(\rho)}_{T_\rho K} \overset{\perp}{\oplus} \underbrace{E_u^{(2)}(\rho) \oplus E_s^{(2)}(\rho)}_{(T_\rho K)^\perp}$$

$$D\Phi \downarrow \qquad\qquad \downarrow \qquad\qquad\qquad \downarrow$$

$$T^*\mathbb{R}^{2d}_{(q,p)} = \underbrace{\left(\mathbb{R}^d_{v_q} \oplus \mathbb{R}^d_{v_p}\right)}_{T^*\mathbb{R}^d_{v_q}} \overset{\perp}{\oplus} \underbrace{\left(\mathbb{R}^d_{\zeta_p} \oplus \mathbb{R}^d_{\zeta_q}\right)}_{T^*\mathbb{R}^d_{\zeta_p}}.$$

See Fig. 2.20. With respect to these coordinates the differential of the canonical map $DF_\rho : T_\rho\left(T^*M\right) \to T_{F(\rho)}\left(T^*M\right)$ is expressed as

$$D\Phi \circ DF_\rho \circ D\Phi^{-1} = F^{(1)} \oplus F^{(2)}, \quad F^{(1)} \equiv \begin{pmatrix} A_x & 0 \\ 0 & {}^tA_x^{-1} \end{pmatrix}, \quad F^{(2)} \equiv \begin{pmatrix} A_x & 0 \\ 0 & {}^tA_x^{-1} \end{pmatrix} \tag{2.88}$$

where

$$A_x \equiv Df\mid_{E_u(x)}: \mathbb{R}^d \to \mathbb{R}^d \tag{2.89}$$

is an expanding linear map. $\|A_x\|_{\min} \geq \lambda > 1$.

At the level of operators, we perform a decomposition similar to (2.87) and obtain a microlocal decomposition of the transfer operator $\hat{F}_N$ as a tensor product $\hat{F}_{N|T_\rho K} \otimes \hat{F}_{N|(T_\rho K)^\perp}$. Precisely we obtain correspondingly to (2.88) above

$$\hat{F}_N \equiv e^V \cdot \mathscr{L}_A \otimes \mathscr{L}_{{}^tA^{-1}} \tag{2.90}$$

with

$$\mathscr{L}_A u := u \circ A^{-1} \text{ on } C_0^\infty\left(\mathbb{R}^d\right)$$

$$\mathscr{L}_{{}^t A^{-1}} u := u \circ {}^t A \text{ on } C_0^\infty\left(\mathbb{R}^d\right).$$

We observe that:

- $|\det A|^{-1/2} \mathscr{L}_A$ is unitary on $L^2\left(\mathbb{R}^d\right)$.
- From model in Theorem 2.5, we have shown that in an anisotropic Sobolev space, $\mathscr{L}_A$ has discrete Ruelle spectrum in bands indexed by $k \geq 0$ and given by:

$$\|A\|_{\max}^{-k} \leq |z_k| \leq \|A\|_{\min}^{-k}$$

and that corresponding eigenspace are homogeneous polynomials of degree k. We observe that the adjoint operator is $\mathscr{L}_A^* = |\det A| . \mathscr{L}_{A^{-1}}$. The spectrum of $\mathscr{L}_A^*$ is the conjugate of that of $\mathscr{L}_A$. We have $\mathscr{L}_{{}^t A^{-1}} = \frac{1}{|\det A|} \cdot \mathscr{L}_A^*$ and deduce that $\mathscr{L}_{{}^t A^{-1}}$ has a discrete Ruelle spectrum in bands indexed by $k \geq 0$ and given by:

$$|\det A|^{-1} \cdot \|A\|_{\max}^{-k} \leq |z_k| \leq |\det A|^{-1} \cdot \|A\|_{\min}^{-k}. \tag{2.91}$$

Therefore we prefer to write (2.90) as

$$\hat{F}_N = e^V \cdot \left(\underbrace{|\det A|^{-1/2} \cdot \mathscr{L}_A}_{\text{unitary}} \right) \otimes \left(\underbrace{|\det A|^{1/2} \cdot \mathscr{L}_{{}^t A^{-1}}}_{\text{discrete spectrum}} \right)$$

and from (2.91) the discrete spectrum of $|\det A|^{1/2} \cdot \mathscr{L}_{{}^t A^{-1}}$ is

$$|\det A|^{-1/2} \cdot \|A\|_{\max}^{-k} \leq |z_k| \leq |\det A|^{-1/2} \cdot \|A\|_{\min}^{-k}.$$

From this microlocal description we obtain that for a given k (this will correspond to the k-th band), the transfer operator $\hat{F}_N$ has "local norm max/min" bounded by

$$e^{\Gamma_k^\pm(x)} = e^V \cdot |\det A|^{-1/2} \cdot \|A\|_{\max/\min}^{-k}.$$

From (2.89) and (2.65) this gives

$$\Gamma_k^\pm(x) = V + \log|\det A|^{-1/2} - k \log \|A\|_{\max/\min}$$

$$= V - \frac{1}{2}\log\left|\det Df_x|_{E_u(x)}\right| - k \log \left\|Df_{|E_u(x)}\right\|_{\max/\min}$$

$$= D(x) - k \log \left\|Df_{|E_u(x)}\right\|_{\max/\min}. \tag{2.92}$$

For the operator $\hat{F}_N^n$ we have similarly that it has "local norm max/min" bounded by $e^{\Gamma_k^{\pm}(x,n)}$ with

$$\Gamma_k^{\pm}(x,n) = D_n(x) - k \log \left\| Df^n_{|E_u(x)} \right\|_{\max/\min}. \tag{2.93}$$

From the previous local description, we can construct explicitly some approximate local spectral projectors Π_k for every value of k, and patching these locals expression together we get global spectral operators for each band (under pitching conditions). We deduce that the spectrum is contained in bands $\mathbf{B}_k$ limited by $\gamma_k^- \leq \log|z| \leq \gamma_k^+$ (image of the projector Π_k) with

$$\gamma_k^+ = \limsup_{n\to\infty} \left(\sup_x \frac{1}{n} \Gamma_k^+(x,n) \right), \quad \gamma_k^+ = \liminf_{n\to\infty} \left(\inf_x \frac{1}{n} \Gamma_k^-(x,n) \right).$$

Then (2.93) gives expressions (2.68) of the Theorem.

The proof of the Weyl law is similar to the proof of J.Sjöstrand about the damped wave equation [40] but needs more arguments. The accumulation of resonances on the value $\exp\langle D\rangle$ uses the ergodicity property and is also similar to the spectral results obtained in [40] for the damped wave equation.

In [23] the proof needs more arguments because one has to show that non linear corrections are negligible.

2.3.4 Ruelle Spectrum for Anosov Vector Fields

We suppose that X is an Anosov vector field on a smooth closed manifold M. Let $V \in C^\infty(M)$ be a smooth function called "*potential function*".

Definition 2.9. The *transfer operator* is the group of operators

$$\hat{F}_t : \begin{cases} C^\infty(M) & \to C^\infty(M) \\ v & \to e^{tA}v \end{cases}, \quad t \geq 0$$

with the generator

$$A := -X + V \tag{2.94}$$

which is a first order differential operator (in local coordinates we have: $A = -\sum_j X^j \frac{\partial}{\partial x^j} + V(x)$).

Remark 2.42.

- Since X generates the flow ϕ_t we can write[16] $\hat{F}_t v = \left(e^{\int_0^t V \circ \phi_{-s} ds} \right) v \left(\phi_{-t}(x) \right)$, hence $\hat{F}_t$ acts as transport of functions by the flow with multiplication by exponential of the function V averaged along the trajectory.
- In the case $V = 0$, the operator $\hat{F}_t$ is useful in order to express "time correlation functions" between $u, v \in C^\infty(M)$, $t \in \mathbb{R}$:

$$C_{u,v}(t) := \int_M u \cdot (v \circ \phi_{-t}) \, dx = \langle u, \hat{F}_t v \rangle_{L^2}. \tag{2.95}$$

The study of these time correlation functions permits to establish the mixing properties and other statistical properties of the dynamics of the Anosov flow.

- In the particular case $V = 0$, $u = cste$ is an obvious eigenfunction of $A = -X$ with eigenvalue $z_0 = 0$.
- If dx is a smooth measure preserved by the flow (this is the case for a contact Anosov flow) then $\operatorname{div} X = 0$ and in the case $V = 0$, we have that $\hat{F}_t$ is unitary in $L^2(M, dx)$ and $iA = (iA)^*$ is self-adjoint and has essential spectrum on the imaginary axis $\operatorname{Re} z = 0$, that is useless. In the next theorem we consider more interesting functional spaces where the operator A has discrete spectrum but is non self-adjoint.

By duality, we extend $A : C^\infty(M) \to C^\infty(M)$ to $A : \mathscr{D}'(M) \to \mathscr{D}'(M)$.

Theorem 2.9 ([11,22] *Discrete Spectrum*, Fig. 2.21). *If X is an Anosov vector field and $V \in C^\infty(M)$ then for every $C > 0$, there exists a Hilbert space $\mathscr{H}_C$ called "anisotropic Sobolev space" with $C^\infty(M) \subset \mathscr{H}_C \subset \mathscr{D}'(M)$, such that*

$$A = -X + V \qquad : \mathscr{H}_C \to \mathscr{H}_C$$

has discrete spectrum on the domain $\operatorname{Re}(z) > -C\lambda$, *called* Ruelle-Pollicott *resonances, independent on the choice of $\mathscr{H}_C$.*

We have an upper bound for the density of resonances: for every $\beta > 0$, in the limit $b \to +\infty$ we have

$$\sharp\{z \in \operatorname{Res}(A), \ |\operatorname{Im}(z) - b| \leq \sqrt{b}, \ \operatorname{Re}(z) > -\beta\} \leq o(b^{n-1/2}), \tag{2.96}$$

with $n = \dim M$.

Remark 2.43. Concerning the meaning of these eigenvalues, notice that with the choice $V = 0$, if $(-X)v = zv$, v is an invariant distribution with eigenvalue

---

[16]To prove this we derive the right hand side $B(x,t) = \left(e^{\int_0^t V \circ \phi_{-s} ds} \right) v \left(\phi_{-t}(x) \right)$, giving $\frac{\partial B}{\partial t} = (V - X) B = AB$. On the other hand $\frac{\partial}{\partial t} \left(\hat{F}_t v \right) = A \left(\hat{F}_t v \right)$ also. Unicity of the solution gives that $B = \hat{F}_t v$.

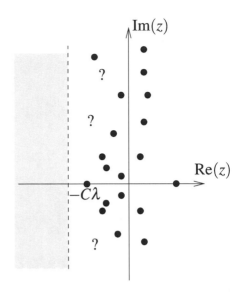

Fig. 2.21 Illustration of Theorem 2.9. The spectrum of $A = -X + V$ is discrete on Re $(z) > -C\lambda$ in space $\mathcal{H}_C$ (for any $C > 0$) but it does not give existence of eigenvalues

$z = -a + ib \in \mathbb{C}$, then $v \circ \phi_{-t} = e^{-tX}v = e^{-at}e^{ibt}v$, i.e. $a = -\text{Re}(z)$ contributes as a damping factor and $b = \text{Im}(z)$ as a frequency in time correlation function (2.95). See Theorem 2.11 below for a precise statement. Notice also the symmetry of the spectrum under complex conjugation that $Av = zv$ implies $A\bar{v} = \bar{z}\bar{v}$.

Remark 2.44. ★ The term *"resonance"* comes from quantum physics where an (elementary or composed) particle usually decay towards other particles. It is modeled by a "resonance", i.e. a quantum state which an eigenvector of the Hamiltonian operator and an eigenvalue $z = -a + ib \in \mathbb{C}$ which behaves as $e^{zt} = e^{-at}e^{ibt}$. The imaginary part of z is written $b = \frac{E}{\hbar}$ with the energy $E = mc^2$ related to the mass $m = \frac{\hbar}{c^2}b$ of the particle. The real part gives $e^{-at} = e^{-t/\tau}$ with $\tau = 1/a$ the "mean life time" of the particle. For example the neutron has $\tau \simeq 15$ mn (very long) and $E = 940$ GeV. In nuclear physics, the mean life time of resonances τ is usually of order 10^{-22} s.

★ See on a movie (http://www-fourier.ujf-grenoble.fr/~faure/articles): the spectrum of the partially expanding map

$$(x, y) \rightarrow (2x \mod 1, y + \sin 2\pi x) \in S^1 \times \mathbb{R}.$$

In Theorem 2.9 the last result gives an upper bound for the number of resonances. The difficulty of giving a lower bound is common in problems which involves "non normal operators" [43] (here A is non normal in $\mathcal{H}_C$). This is due to the fact that for non normal operators, the spectrum may be very unstable with respect to perturbation. The simplest example to have in mind is the following $N \times N$ matrix with parameter $\varepsilon \in \mathbb{R}$:

$$M_\varepsilon = \begin{pmatrix} 0 & 1 & 0 & 0 \\ 0 & 0 & & \ddots \\ & & \ddots & 1 \\ \varepsilon & & & 0 \end{pmatrix}.$$

For $\varepsilon = 0$ the spectrum is 0 with multiplicity N. For $\varepsilon > 0$ is it easy to check that there are N eigenvalues on the circle of radius $r_{\varepsilon,N} = \varepsilon^{1/N}$. So for $\varepsilon = 10^{-10}$, and $N = 10$ the radius is $r = 0.1$.

2.3.4.1 Sketch of Proof of Theorem 2.9

This proof (taken from [22]) uses semiclassical analysis. Let us consider the differential operator

$$P := iA \underset{(2.94)}{=} -iX + iV. \qquad (2.97)$$

On the cotangent space T^*M we denote $x \in M$ and $\xi \in T^*_x M$. The principal symbol of P is the function $p \in C^\infty (T^*M)$ given by (see (2.131) or [42, p. 2])

$$p(x, \xi) = X_x(\xi). \qquad (2.98)$$

The function p defines a Hamiltonian vector field $\mathbf{X}$ on T^*M by $\Omega(\mathbf{X}, .) = dp$, with $\Omega = \sum_j dx^j \wedge d\xi^j$ being the canonical symplectic form. In fact $\mathbf{X}$ is the canonical lift of X on the cotangent space. Its flow

$$\Phi_t = e^{-t\mathbf{X}} \qquad (2.99)$$

is a lift of $\phi_t : M \to M$ and acts lineary in the fibers $\Phi_t : T^*_x M \to T^*_{\phi_t(x)} M$. It preserves the decomposition of the cotangent bundle

$$T^*_x M = E^*_u(x) \oplus E^*_s(x) \oplus E^*_0(x)$$

defined as the dual decomposition of the tangent space (2.14) by

$$E^*_u(E_u \oplus E_0) = 0, \quad E^*_s(E_s \oplus E_0) = 0, \quad E^*_0(E_u \oplus E_s) = 0.$$

From Eq. (2.16), we have that $E^*_0 = \mathbb{R}\alpha$. For a point $(x, \xi) \in T^*M$ we can consider $\mathscr{E} = p(x, \xi) = X_x(\xi)$ as the component of ξ along the axis $E^*_0(x)$, called the energy and preserved by the flow. The energy level is $\Sigma_\mathscr{E} := p^{-1}(\mathscr{E})$. From (2.98) and (2.16), $\Sigma_\mathscr{E}$ is an affine subbundle of T^*M given by

$$\Sigma_\mathscr{E} = p^{-1}(\mathscr{E}) = (\mathscr{E} \cdot \alpha) + \left(E^*_u \oplus E^*_s\right).$$

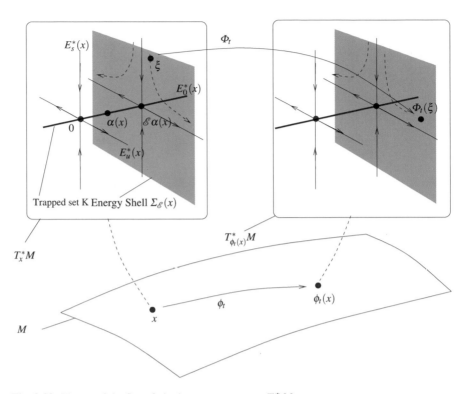

Fig. 2.22 Picture of the flow Φ_t in the cotangent space T^*M

By duality, for $t > 0$, the map $\Phi_t : E_u^*(x) \rightarrow E_u^*(\phi_t(x))$ is expanding and $\Phi_t : E_s^*(x) \rightarrow E_s^*(\phi_t(x))$ is contracting. See Fig. 2.22.

The *trapped set* (or non wandering set) of the flow Φ_t is defined as the set of point who do not escape to infinity in the past or future:

$$K := \{(x, \xi) \in T^*M, \exists C \Subset T^*M \text{ compact}, \forall t \in \mathbb{R}, \Phi_t(x, \xi) \in C\} \subset T^*M.$$

From the previous description we have that the trapped set is the rank one subbundle E_0^*: $K = E_0^*$, $\dim K = \dim M + 1$.

For an arbitrary large constant $C > 0$, we construct an *escape function* $a(x, \xi)$ on T^*M such that[17] far from the trapped set K one has: $\mathbf{X}(a) \ll -C.\lambda$.

Then let us consider the conjugated operator

$$\tilde{P} := e^{\mathrm{Op}(a)} P e^{-\mathrm{Op}(a)} = P + [\mathrm{Op}(a), P] + \dots . \tag{2.100}$$

---

[17]Precisely we choose $e^{a(x,\xi)} = \langle \xi \rangle^{m(x,\xi)}$ i.e. $a(x, \xi) = m(x, \xi) \log \langle \xi \rangle$ with $m(x, \xi) = \pm C$ along the stable/unstable directions $E_{s,u}^*(x)$ respectively. Hyperbolicity assumption gives that $\mathbf{X}(\xi_{s/u}) = \mp \lambda.\xi_{s/u}$ hence $\mathbf{X}(a) = m.\mathbf{X}(\log(\xi_{s/u})) = -C\lambda$.

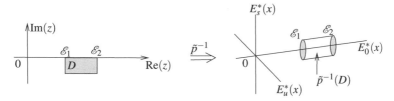

Fig. 2.23 As $\tilde{p}^{-1}(D) \subset T^*M$ is compact, $\tilde{P} = \mathrm{Op}(\tilde{p})$ has discrete spectrum on D

From (2.17) its symbol is

$$
\begin{aligned}
\tilde{p}(x,\xi) &= p(x,\xi) - i\{a, p\} + iV + O\left(S^{-1+0}\right) \\
&= X(\xi) + i\mathbf{X}(a) + iV + O\left(S^{-1+0}\right).
\end{aligned}
\tag{2.101}
$$

Let $D \subset \mathbb{C}$ a compact domain of the spectral plane. If $C > 0$ is large enough then $\tilde{p}^{-1}(D)$ is a compact subset of T^*M. See Fig. 2.23.

As a consequence

$$
\tilde{P} : L^2(M) \to L^2(M)
\tag{2.102}
$$

has discrete spectrum[18] on the domain D.

Let

$$
\mathscr{H}_C := e^{-\mathrm{Op}(a)} L^2(M)
$$

be the *anisotropic Sobolev space*. Equivalently, from (2.100) and (2.97), (2.102) gives

$$
P \quad : \mathscr{H}_C \to \mathscr{H}_C, \qquad A = -iP \quad : \mathscr{H}_C \to \mathscr{H}_C
$$

have discrete spectrum respectively on the domain D and $-iD$.

The Weyl upper bound is obtained by computing the symplectic volume of $\tilde{p}^{-1}(D)$.

---

[18]To show the general statement used here that $\tilde{p}^{-1}(D) \subset T^*M$ is compact implies that $\tilde{P} = \mathrm{Op}(\tilde{p}) : L^2(M) \to L^2(M)$ has discrete spectrum on D we use the resolvent as follows: let $z_0 \in D$. From "*semiclassical functional calculus*"[17, 30], $R_{\tilde{p}}(z_0) := (z_0 - \tilde{P})^{-1}$ is a PDO with symbol $r_{\tilde{p}}(z_0) = (z_0 - \tilde{p})^{-1}$. From (2.101) on can write

$$
r_{\tilde{p}}(z_0) = r_{\tilde{p}-K}(z_0) + r_{\tilde{p}-K}(z_0) K r_{\tilde{p}}(z_0)
$$

where the first term of the right is bounded so that $\mathrm{Op}\left(r_{\tilde{p}-K}(z_0)\right)$ has a small norm and the second term decay so that $\mathrm{Op}(//)$ is compact. With this kind of argument, we deduce that $\tilde{P}$ has discrete spectrum in D.

2.3.5 Ruelle Band Spectrum for Contact Anosov Vector Fields

We present here the result announced in [24].

Remark 2.45. Recently there appeared few papers where the authors obtain results for contact Anosov flows using this semiclassical approach: spectral gap estimate and decay of correlation [38], Weyl law upper bound [13] and meromorphic properties of the dynamical zeta function [19]. We would like to mention also a closely related work: in [18], for a problem concerning decay of waves around black holes, S. Dyatlov show that the spectrum of resonances has a band structure similar to what is observed for contact Anosov flows. In fact these two problems are very similar in the sense that in both cases the trapped set is symplectic and normally hyperbolic. This geometric property is the main reason for the existence of a band structure. However in [18], some regularity of the hyperbolic foliation is required and that regularity is not present for contact Anosov flows.

2.3.5.1   Case of Geodesic Flow on Constant Curvature Surface

In Sect. 2.5 we have observed that there is a contact Anosov flow X on $\Gamma \backslash SL_2\mathbb{R}$ corresponding to the geodesic flow on $\Gamma \backslash \mathbb{H}^2$.

Using representation theory, it is known that the Ruelle-Pollicott spectrum of the operator $(-X)$ coincides with the zeros of the dynamical Fredholm determinant. This dynamical Fredholm determinant is expressed as the product of the Selberg zeta functions and gives the following result; see Fig. 2.25a. We refer to [25] for further details.

Proposition 2.8. *If X is the geodesic flow on an hyperbolic surface $\mathscr{S} = \Gamma \backslash \mathbb{H}^2$ then the Ruelle-Pollicott eigenvalues z of $(-X)$, i.e. giving $(-X)u = z \cdot u$ with $u \in \mathscr{H}_C$, are of the form*

$$z_{k,l} = -\frac{1}{2} - k \pm i\sqrt{\mu_l - \frac{1}{4}} \tag{2.103}$$

where $k \in \mathbb{N}$ and $(\mu_l)_{l \in \mathbb{N}} \in \mathbb{R}^+$ are the discrete eigenvalues of the hyperbolic Laplacian $\Delta = -y^2 \left(\frac{\partial^2}{\partial x^2} + \frac{\partial^2}{\partial y^2} \right)$ on the surface $\mathscr{S} = \Gamma \backslash \mathbb{H}^2$. There are also $z_n = -n$ with $n \in \mathbb{N}^*$. Each set $(z_{k,l})_l$ with fixed k will be called the line $\mathbf{B}_k$. The "Weyl law" for Δ gives the density of eigenvalues on each vertical line $\mathbf{B}_k$, for $b \to \infty$,

$$\sharp\{z_{k,l}, \ b < \mathrm{Im}\,(z_{k,l}) < b + 1\} \sim |b|\frac{\mathscr{A}}{2\pi} \tag{2.104}$$

where $\mathscr{A}$ is the area of $\mathscr{S}$.

Proof. For the proof we can use representation theory: it is known that the Ruelle-Pollicott spectrum of the operator $(-X)$ coincides with the zeros of the dynamical Fredholm determinant. This dynamical Fredholm determinant is expressed as the product of the Selberg zeta functions.

Here is an argument that Ruelle resonances are related to the spectrum of the Laplacian and comes by bands. Suppose that $(-X) u = zu$ is a Ruelle-Pollicott eigenvector. From (2.19) we deduce that:

$$(-X) (Uu) = (-UX + U) u = (z + 1) (Uu),$$

$$(-X) (Su) = (-SX - S) u = (z - 1) (Su).$$

This gives a family of other eigenvalues $z + k$, $k \in \mathbb{Z}$. But the condition that the spectrum is in the domain $\mathrm{Re}(z) \leq 0$ implies that there exists $k \geq 1$ such that $U^k u = 0, U^{k-1} u \neq 0$. We say that $u \in \mathbf{B}_k$ belongs to the band k. Notice also that if $u \in \mathbf{B}_0$ i.e. $Uu = 0$ then using the *Casimir operator* $\Delta = -X^2 - \frac{1}{2} SU - \frac{1}{2} US$ of $SL_2 \mathbb{R}$ we have

$$\Delta u = \left(-X^2 - \frac{1}{2} SU - \frac{1}{2} US \right) u \underset{(2.19)}{=} \left(-X^2 + X - SU \right) u = -z (z + 1) u = \mu u.$$

Let $\langle u \rangle_{SO_2} \in \mathscr{D}'(\mathscr{M})$ be the distribution u averaged by the action of SO_2. We suppose that $\langle u \rangle_{SO_2} \neq 0$. It is shown in [22] that the wavefront of u is included in the unstable manifold $E_u^* \subset T^* M$. Using an argument of Hörmander, since E_u^* is not contained in the kernel of $\Theta = S - U$ the generator of SO_2, then this wavefront is killed by the action of SO_2 and $\langle u \rangle_{SO_2} \in C^\infty(\mathscr{M})$ is in fact a smooth function on the surface $\mathscr{M} = \Gamma \backslash (SL_2\mathbb{R}/SO_2)$. Moreover since Δ commutes with the action of SO_2, we still have that $\Delta \langle u \rangle_{SO_2} = \mu \langle u \rangle_{SO_2}$ with $\Delta \equiv -y^2 \left(\frac{\partial^2}{\partial x^2} + \frac{\partial^2}{\partial y^2} \right)$ being the hyperbolic Laplacian. Δ being elliptic on $\mathscr{M}$ also implies that $\langle u \rangle_{SO_2}$ is smooth. From spectral theory in $L^2(\mathscr{M})$, Δ is a positive self-adjoint operator and has discrete real and positive eigenvalues $\mu_l = -z(z+1) \geq 0$. Therefore the Ruelle eigenvalue is

$$z = -\frac{1}{2} \pm i \sqrt{\mu_l - \frac{1}{4}}.$$

We deduce the other Ruelle eigenvalues by the shift $z - k$, $k \in \mathbb{N}$. The Weyl law for the Laplacian gives (2.104). Using Representation theory we can show that they are no other eigenvalues [27]; i.e. that $\mathrm{Im} z \neq 0$ implies that $\langle u \rangle_{SO_2} \neq 0$ See Fig. 2.24.

2.3.5.2 General Case

Proposition 2.8 above shows that the Ruelle-Pollicott spectrum for the geodesic flow on constant negative surface has the structure of vertical lines $\mathbf{B}_k$ at

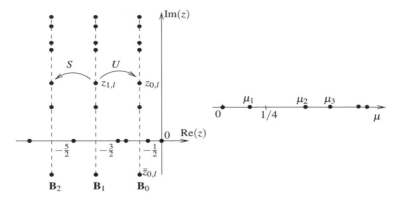

Fig. 2.24 Ruelle Pollicott resonances for the geodesic flow on a hyperbolic surface

$\text{Re} z = -\frac{1}{2} - k$. In each line the eigenvalues are in correspondence with the eigenvalues of the Laplacian Δ. We address now the question if this structure persists somehow for geodesic flow on manifolds with negative (variable) sectional curvature and more generally for any contact Anosov flow.

We consider here an contact Anosov vector field X on a smooth closed manifold M and a smooth potential function $V \in C^{\infty}(M)$.

Remark 2.46. "Concerning the leading eigenvalue". Similarly to (2.59) above, we can show that for contact Anosov flow the Ruelle spectrum has a leading real eigenvalue $z_0 \in \mathbb{R}$ (i.e. other eigenvalues are $\text{Re}(z_j) < z_0$) given by

$$z_0 = \text{Pr}(V - J)$$

where $J = \text{div} X_{|E_u}$ is the "*unstable Jacobian*"[19] and for a function $\varphi \in C(M)$,

$$\text{Pr}(\varphi) := \lim_{t \to \infty} \frac{1}{t} \log \left(\sum_{\gamma, |\gamma| \leq t} \exp \left(\int_0^t \varphi \right)(\gamma) \right)$$

is called the *topological pressure*.

---

[19]Let μ_g be the induced Riemann volume form on $E_u(x)$ defined from the choice of a metric g on M. As the usual definition in differential geometry [41, p. 125], for tangent vectors $u_1, \dots u_d \in E_u(x)$, $\text{div} X_{|E_u}$ measures the rate of change of the volume of E_u and is defined by

$$\left(\text{div} X_{|E_u}(x) \right) \cdot \mu_g(u_1, \dots u_d) = \lim_{t \to 0} \frac{1}{t} \left(\mu_g(D\phi_t(u_1), \dots, D\phi_t(u_d)) - \mu_g(u_1, \dots u_d) \right).$$

Equivalently we can write that

$$\text{div} X_{|E_u}(x) = \frac{d}{dt} \left(\det(D\phi_t)_{|E_u} \right)_{t=0}. \tag{2.105}$$

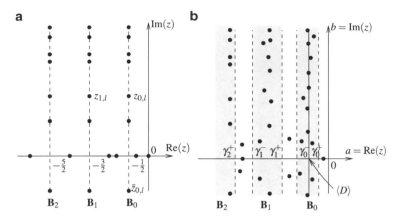

Fig. 2.25 (**a**) For an hyperbolic surface $\mathscr{S} = \Gamma\backslash\mathbb{H}^2$, the Ruelle-Pollicott spectrum of the geodesic vector field $-X$ given by Proposition 2.8. It is related to the eigenvalues of the Laplacian by (2.103); (**b**) For a general contact Anosov flow, the spectrum of $A = -X + V$ and its asymptotic band structure given by Theorems 2.10

We introduce now the following function called *"potential of reference"* that will play an important role

$$V_0(x) := \frac{1}{2}\mathrm{div}\,X_{|E_u} = \frac{1}{2}J. \tag{2.106}$$

Remark 2.47. From (2.15) we have $V_0(x) \geq \frac{1}{2}d \cdot \gamma$. Since $E_u(x)$ is only Hölder in x so is $V_0(x)$.

We will also consider the difference

$$D(x) := V(x) - V_0(x) \tag{2.107}$$

and called it the *"effective damping function"*. For simplicity we will write:

$$\left(\int_0^t D\right)(x) := \int_0^t (D \circ \phi_{-s})(x)\,ds, \quad x \in M,$$

for the *Birkhoff sum* of D along trajectories. Finally we recall the notation $\|L\|_{min,max}$ in (2.42) for an invertible linear operator. The following theorem is similar to Theorem 2.7 that was for prequantum maps.

Theorem 2.10 ([25] Asymptotic band structure, Fig. 2.25). *If X is a contact Anosov vector field on M and $V \in C^\infty(M)$ then for every $C > 0$, there exists an Hilbert space $\mathscr{H}_C$ with $C^\infty(M) \subset \mathscr{H}_C \subset \mathscr{D}'(M)$, such that for any $\varepsilon > 0$, the Ruelle-Pollicott eigenvalues $(z_j)_j \in \mathbb{C}$ of the operator $A = -X + V : \mathscr{H}_C \to \mathscr{H}_C$ on the domain $\mathrm{Re}(z) > -C\lambda$ are contained, up to finitely many exceptions, in the* *union of finitely many bands*

$$z \in \bigcup_{k \geq 0} \underbrace{\left[\gamma_k^- - \varepsilon, \gamma_k^+ + \varepsilon \right] \times i\mathbb{R}}_{\text{Band } \mathbf{B}_k}$$

with for $k \geq 0$,

$$\gamma_k^+ = \lim_{t \to \infty} \left| \sup_x \frac{1}{t} \left(\left(\int_0^t D \right)(x) - k \log \left\| D\phi_t(x)_{/E_u} \right\|_{min} \right) \right|, \qquad (2.108)$$

$$\gamma_k^- = \lim_{t \to \infty} \left| \inf_x \frac{1}{t} \left(\left(\int_0^t D \right)(x) - k \log \left\| D\phi_t(x)_{/E_u} \right\|_{max} \right) \right| \qquad (2.109)$$

and where $D = V - V_0$ is the damping function (2.107). In the gaps (i.e. between the bands) the norm of the resolvent is controlled: there exists $c > 0$ such that for every $z \notin \bigcup_{k \geq 0} \mathbf{B}_k$ with $|\text{Im}(z)| > c$

$$\left\| (z - A)^{-1} \right\| \leq c. \qquad (2.110)$$

For some $k \geq 0$, if the band $\mathbf{B}_k$ is "isolated", i.e. $\gamma_{k+1}^+ < \gamma_k^-$ and $\gamma_k^+ < \gamma_{k-1}^-$ (this last condition is for $k \geq 1$) then the number of resonances in $\mathbf{B}_k$ obeys a "Weyl law": $\forall b > c$,

$$\frac{1}{c} |b|^d < \frac{1}{|b|^\varepsilon} \cdot \sharp \left\{ z_j \in \mathbf{B}_k, b < \text{Im}(z_j) < b + b^\varepsilon \right\} < c |b|^d \qquad (2.111)$$

with $\dim M = 2d + 1$. The upper bound holds without the condition that $\mathbf{B}_k$ is isolated. If the external band $\mathbf{B}_0$ is isolated i.e. $\gamma_1^+ < \gamma_0^-$, then most of the resonances accumulate on the vertical line

$$\text{Re}(z) = \langle D \rangle := \frac{1}{\text{Vol}(M)} \int_M D(x) \, dx$$

in the precise sense that

$$\frac{1}{\sharp \mathscr{B}_b} \sum_{z_i \in \mathscr{B}_b} |\text{Re}(z_i) - \langle D \rangle| \xrightarrow[b \to \infty]{} 0, \qquad \text{with } \mathscr{B}_b := \{ z_i \in \mathbf{B}_0, |\text{Im}(z_i)| < b \}.$$

$$(2.112)$$

Remark 2.48. In 2009 M. Tsujii has obtained γ_0^+ in [44, 45]. He also obtained the estimate (2.110) for $\text{Re}(z) \geq \gamma_0^+ + \varepsilon$.

Remark 2.49. For a general contact Anosov vector field it is possible to choose the potential $V = V_0$ (although it is non smooth)[25], giving $\gamma_0^+ = \gamma_0^- = 0$, i.e. the first band is reduced to the imaginary axis and is isolated from the second band by a gap, $\gamma_1^+ < 0$, cp. Fig. 2.26.

Fig. 2.26 Ruelle-Pollicott spectrum for a general Contact Anosov flow and with potential $V_0 = \frac{1}{2} \mathrm{div} X_{|E_u}$

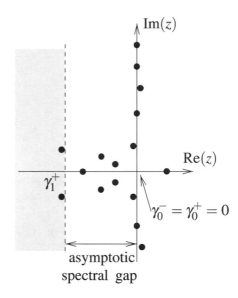

2.3.5.3 Consequence for Correlation Functions Expansion

We mentioned the usefulness of dynamical correlation functions in (2.95). Let Π_j denotes the finite rank spectral projector associated to the eigenvalue z_j. The following Theorem provides an expansion of correlation functions over the spectrum of resonances of the first band $\mathbf{B}_0$. This is an infinite sum.

Theorem 2.11 ([24]). *Suppose that* $\gamma_1^+ < \gamma_0^-$. *Then for any* $\varepsilon > 0, \exists C_\varepsilon > 0$, *any* $u, v \in C^\infty (M)$ *and* $t \geq 0$,

$$\left| \langle u, \hat{F}_t v \rangle_{L^2} - \sum_{z_j, \mathrm{Re}(z_j) \geq \gamma_1^+ + \varepsilon} \langle u, \hat{F}_t \Pi_j v \rangle \right| \leq C_\varepsilon \cdot \|u\|_{\mathscr{H}_C'} \|v\|_{\mathscr{H}_C} e^{(\gamma_1^+ + \varepsilon)t}. \quad (2.113)$$

The infinite sum above converges because for arbitrary large $m \geq 0$ *there exists* $C_{m,\varepsilon}(u, v) \geq 0$ *such that* $\left| \langle u, \hat{F}_t \Pi_j v \rangle \right| \leq C_{m,\varepsilon}(u, v) \cdot \left| \mathrm{Im}(z_j) \right|^{-m} \cdot e^{(\gamma_0^+ + \varepsilon)t}$.

Remark 2.50. Equation (2.113) is a refinement of decay of correlation results of Dolgopyat [14], Liverani [35], Tsujii [44, 45, Cor.1.2] and Nonnenmacher-Zworski [38, Cor.5] where their expansion is a finite sum over one or a finite number of leading resonances.

Remark 2.51. In the case of simple eigenvalues $z_j = -a_j + ib_j$ then Π_j is a rank one projector and $\langle u, \hat{F}_t \Pi_j v \rangle = e^{-a_j t} e^{ib_j t} \langle u, \Pi_j v \rangle$.

Remark 2.52. As we did in (2.74), we call the second term of (2.113), the *quantum correlation function*.

2.3.5.4   Proof of Theorem 2.10

The band structure and all related results presented in Theorem 2.10 have already
been proven for the spectrum of Anosov prequantum map in [23] and presented in
Theorem 2.7. An Anosov prequantum map $\tilde{f} : P \to P$ is an equivariant lift of an
Anosov diffeomorphism $f : M \to M$ on a principal bundle $U(1) \to P \to M$
such that $\tilde{f}$ preserves a contact one form α (a connection on P). Therefore $\tilde{f} : P \to
P$ is very similar to the contact Anosov flow $\phi_t : M \to M$ considered here, that
also preserves a contact one form α. Our proof of Theorem 2.10 is directly adapted
from the proof given in [23] and presented in Sect. 2.3.3.1. We refer to this paper for
more precisions on the proof and we use the same notations below. The techniques
rely on semiclassical analysis adapted to the geometry of the contact Anosov flow
lifted in the cotangent space T^*M. In the limit $|\text{Im}z| \to \infty$ of large frequencies
under study, the semiclassical parameter is written $\hbar := 1/|\text{Im}z|$. We now sketch
the main steps of the proof.

The proof is very similar to that of Theorem 2.7. Recall that $\dim M = 2d + 1$, so
$\dim T^*M = 2(2d + 1)$.

Global Geometrical Description

$A = -X + V$ is a differential operator. Its principal symbol is the function
$\sigma(A)(x, \xi) = X_x(\xi)$ on phase space T^*M (the cotangent bundle). It generates
an Hamiltonian flow which is simply the canonical lift of the flow ϕ_t on M. See
Fig. 2.22. Due to Anosov hypothesis on the flow, the non-wandering set of the
Hamiltonian flow is the continuous sub-bundle $K = \mathbb{R}\alpha \subset T^*M$ where α is the
Anosov one form. K is *normally hyperbolic*. This analysis has already been used
in [22] for the semiclassical analysis of Anosov flow (not necessary contact). With
the additional hypothesis that α is a smooth contact one form, the following Lemma
shows that the trapped set $K\backslash\{0\}$ is a smooth symplectic submanifold of T^*M
(usually called the symplectization of the contact one form α).

Lemma 2.5 ([2]). *The trapped set $K\backslash\{0\} = (\mathbb{R}\alpha)\backslash\{0\}$ is a symplectic submanifold of T^*M of dimension* $\dim = 2(d + 1)$, *called the* symplectization *of the
contact one form α.*

Proof. Denote $\pi : T^*M \to M$ the projection map. A point on the trapped set
$K \subset T^*M$ can be written $\xi = \mu \cdot \alpha(x)$ with $\mu \in \mathbb{R}$ and $x \in M$. The Liouville one
form on T^*M at point $\xi = \mu \cdot \alpha(x) \in K$ is

$$\sum_{j=1}^{2d+1} \xi^j dx^j \equiv \mu \cdot \pi^*(\alpha).$$

For simplicity we write the previous equation $\xi dx = \mu \cdot \alpha$.

Then

$$d\,(\xi dx) = d\,(\mu\alpha) = d\mu \wedge \alpha + \mu d\alpha$$

is a 2 form on K giving the following volume form on $K \backslash \{0\}$:

$$(d\,(\mu\alpha))^{d+1} = (d+1)\,\mu^d \cdot d\mu \wedge \alpha \wedge (d\alpha)^d$$

which is non degenerate on $K \backslash \{0\}$ since $\alpha \wedge (d\alpha)^d$ is supposed to be non degenerated on M. In other words the canonical two form $\Omega = \sum_{j=1}^{2d+1} dx^j \wedge d\xi^j = -d\,(\xi dx)$ restricted to $K \backslash \{0\}$ is symplectic.

Let $\rho = (x, \xi) \in K$ be a point on the trapped set. Let $\hbar^{-1} = X_x(\xi) = \mathscr{E}$ be its "energy". Let $\Omega = \sum_j dx^j \wedge d\xi^j$ be the canonical symplectic form on T^*M and consider the Ω-orthogonal splitting of the tangent space at $\rho \in K$:

$$T_\rho\,(T^*M) = T_\rho K \overset{\perp_\Omega}{\bigoplus} (T_\rho K)^{\perp_\Omega}. \tag{2.114}$$

Due to hyperbolicity assumption, we have an additional decomposition of the space

$$(T_\rho K)^{\perp} = \underbrace{E_u^{(2)} \oplus E_s^{(2)}}_{2d}$$

transverse to the trapped set into unstable/stable spaces, i.e. $E_u^{(2)} := (T_\rho K)^{\perp} \cap E_u^*(\rho)$ etc. We have written the dimension below. Also tangent to the trapped set K we have

$$T_\rho K = \underbrace{\left(\underbrace{E_u^{(1)} \oplus E_s^{(1)}}_{2d}\right)}_{} \overset{\perp_\Omega}{\bigoplus} \left(\underbrace{E_0 \oplus E_0^*}_{2}\right).$$

Correspondingly the differential of the lifted flow (2.99), $\Phi_t = e^{-t\mathbf{X}} : T^*M \to T^*M$ is decomposed as

$$D\Phi_t \equiv D\Phi_t^{(1)} \overset{\perp_\Omega}{\bigoplus} D\Phi_t^{(0)} \overset{\perp_\Omega}{\bigoplus} D\Phi_t^{(2)} \tag{2.115}$$

with

$$D\Phi_t^{(1)} \equiv D\Phi_t^{(2)} \equiv \begin{pmatrix} L_x & 0 \\ 0 & {}^t L_x^{-1} \end{pmatrix} : \mathbb{R}^{2d} \to \mathbb{R}^{2d}$$

is linear symplectic with

$$L_x := (D\phi_t)_{|E_u(x)} \tag{2.116}$$

being a linear expanding map. We have $\|L_x\|_{\min} > e^{t\gamma} > 1$ and $D\Phi_t^{(0)} = \mathrm{Id}_{|\mathbb{R}^2}$.

Partition of Unity

We choose an energy $\mathscr{E} = \frac{1}{\hbar} \gg 1$. We decompose functions on the manifold using a microlocal partition of unity of size $\hbar^{1/2-\varepsilon}$ with some $1/2 > \varepsilon > 0$, that is refined as $\hbar \to 0$. In each chart we use a canonical change of variables adapted to the decomposition (2.114) and construct an escape function adapted to the local splitting $E_u^{(2)} \oplus E_s^{(2)}$ above. This escape function has "strong damping effect" outside a vicinity of size $O\left(\hbar^{1/2}\right)$ of the trapped set K. We use this to define the anisotropic Sobolev space $\mathscr{H}_C$. At the level of operators, we perform a decomposition similar to (2.114) and obtain a microlocal decomposition of the transfer operator $\hat{F}_t = e^{tA}$ as a tensor product $\hat{F}_{t|T_\rho K} \otimes \hat{F}_{t|(T_\rho K)^\perp}$. Precisely we obtain correspondingly to (2.115) above

$$\hat{F}_t = e^{tA} \underset{\text{microloc.}}{\equiv} e^{\int_0^t V} \cdot \mathscr{L}_L \otimes e^{-i\mathscr{E}t} \mathrm{Id}_{\mathbb{R}} \otimes \mathscr{L}_{t\,L^{-1}} \tag{2.117}$$

with

$$\mathscr{L}_L u := u \circ L^{-1} \text{ on } C_0^\infty\left(\mathbb{R}^d\right)$$

$$\mathscr{L}_{t\,L^{-1}} u := u \circ^t L \text{ on } C_0^\infty\left(\mathbb{R}^d\right)$$

and $\underset{\text{microloc.}}{\equiv}$ means after multiplication of some cutoff function defining a partition of unity, and up to conjugation by some unitary (Fourier integral operators, F.I.O) operators. We observe that:

- $|\det L|^{-1/2} \mathscr{L}_L$ is unitary on $L^2\left(\mathbb{R}^d\right)$.
- From model in Theorem 2.5, we have shown that in some anisotropic Sobolev space, $\mathscr{L}_L$ has discrete Ruelle spectrum in bands indexed by $k \geq 0$ and that:

$$C_0^{-1} \|L\|_{\max}^{-k} \leq \frac{\|\mathscr{L}_L u\|_{\mathscr{H}_C}}{\|u\|_{\mathscr{H}_C}} \leq C_0 \|L\|_{\min}^{-k} \tag{2.118}$$

and the corresponding group of eigenspaces are *homogeneous polynomials* on $\mathbb{R}^d$ of degree k. We observe that the adjoint operator is $\mathscr{L}_L^* = |\det L| . \mathscr{L}_{L^{-1}}$. For the adjoint $\mathscr{L}_L^*$ we have similar bounds. We have $\mathscr{L}_{t\,L^{-1}} = \frac{1}{|\det L|} \cdot \mathscr{L}_{t\,L}^*$ and deduce that $\mathscr{L}_{t\,L^{-1}}$ has also a discrete Ruelle spectrum in bands indexed by $k \geq 0$

and similar bounds but with the additional factor $|\det L|^{-1}$. Therefore we prefer to write (2.117) as

$$
e^{tA} = e^{\int_0^t V} \cdot \left(\underbrace{|\det L|^{-1/2} \cdot \mathscr{L}_L}_{\text{unitary}} \right) \otimes e^{-i\mathscr{E}t} \mathrm{Id}_{\mathbb{R}} \otimes \left(\underbrace{|\det L|^{1/2} \cdot \mathscr{L}_{tL^{-1}}}_{\text{discrete bands}} \right)
$$

and from (2.118) the discrete spectrum of $|\det L|^{1/2} \cdot \mathscr{L}_{tL^{-1}}$ is in bands with the bounds

$$
C_0^{-1} |\det L|^{-1/2} \cdot \|L\|_{\max}^{-k} \leq \frac{\| |\det L|^{1/2} \cdot \mathscr{L}_{tL^{-1}} u \|_{\mathscr{H}_C}}{\|u\|_{\mathscr{H}_C}} \leq C_0 |\det L|^{-1/2} \cdot \|L\|_{\min}^{-k} .
$$

From this microlocal description we obtain that for given k, the transfer operator e^{tA} has "local norm max/min" bounded by

$$
e^{\Gamma_k^{\pm}(x,t)} \asymp e^{\int_0^t V} \cdot |\det L|^{-1/2} \cdot \|L\|_{\max/\min}^{-k} .
$$

From (2.116) and (2.65) this gives

$$
\Gamma_k^{\pm}(x,t) = \int_0^t V - \frac{1}{2} \log \left| \det \phi_{t|E_u(x)} \right|^{-1/2} - k \log \left\| D\phi_{t|E_u(x)} \right\|_{\max/\min} + O(1)
$$

$$
= \int_0^t D - k \log \left\| D\phi_{t|E_u(x)} \right\|_{\max/\min} + O(1). \tag{2.119}
$$

From the previous local description, we can construct explicitly some approximate local spectral projectors Π_k for every value of k, and patching these locals expressions together we get global spectral operators for each band (under pitching conditions). For the generator A of e^{tA} we deduce that the spectrum is contained in bands $\mathbf{B}_k$ limited by $\gamma_k^- \leq \mathrm{Re}(z) \leq \gamma_k^+$ (image of the projector Π_k) with

$$
\gamma_k^+ = \limsup_{t \to \infty} \left(\sup_x \frac{1}{t} \Gamma_k^+(x,t) \right), \quad \gamma_k^+ = \liminf_{t \to \infty} \left(\inf_x \frac{1}{t} \Gamma_k^-(x,t) \right).
$$

Then (2.119) gives expressions (2.108) of the Theorem.

The proof of the Weyl law (2.111) is similar to the proof of J.Sjöstrand about the damped wave equation [40] but needs more arguments. The accumulation of resonances on the value $\langle D \rangle$ given by the spatial average of the damping function, Eq. (2.112), uses the ergodicity property of the Anosov flow and is also similar to the spectral results obtained in [40] for the damped wave equation.

2.4 Trace Formula and Zeta Functions

We have already presented the Atiyah Bott trace formula in Sect. 2.3.2.2. This "simple formula" is at the basis for exact relations between the Ruelle spectrum and periodic orbits of the dynamics. We have saw such a relation in (2.55) for Anosov maps.

In this section we want to present more precisely what this relation gives when there is a band structure in the Ruelle spectrum. This is the case for prequantum Anosov maps or contact Anosov flows. A consequence of this will be some refined counting formula for periodic orbits.

2.4.1 Gutzwiller Trace Formula for Anosov Prequantum Map

In this section we consider the prequantum transfer operators $\hat{F}_N$ defined in (2.62). We assume the condition $r_1^+ < r_0^-$. (This condition holds if we consider the potential of reference $V = V_0$) As in (2.72), let $\Pi_\hbar : \mathcal{H}_N^r \to \mathcal{H}_N^r$ be the spectral projector for the external band and let $\mathcal{H}_\hbar$ be its image called quantum space. Let $\hat{\mathcal{F}}_\hbar : \mathcal{H}_\hbar \to \mathcal{H}_\hbar$ be the restriction of $\hat{F}_N$ to $\mathcal{H}_\hbar$.

Theorem 2.12 ([23]"Gutzwiller trace formula for large time"). *Let $\varepsilon > 0$. For any $\hbar = 1/(2\pi N)$ small enough, in the limit $n \to \infty$, we have*

$$\left| Tr\left(\hat{\mathcal{F}}_\hbar^n\right) - \sum_{x=f^n(x)} \frac{e^{D_n(x)} e^{iS_{n,x}/\hbar}}{\sqrt{\left|Det\left(1 - Df_x^n\right)\right|}} \right| < CN^d (r_1^+ + \varepsilon)^n \qquad (2.120)$$

where $e^{i2\pi S_{n,x}}$ is the action of a periodic point defined in (2.13) and D_n is the Birkhoff sum (2.66) of the effective damping function $D(x) = V(x) - V_0(x)$.

2.4.1.1 The Question of Existence of a "Natural Quantization"

The following problem is a recurrent question in mathematics and physics in the field of quantum chaos, since the discovery of the Gutzwiller trace formula. For simplicity of the discussion we consider $V = V_0$ i.e. no effective damping, as in Fig. 2.18.

Problem 2.1. Does there exists a sequence $\hbar_j > 0$, $\hbar_j \to 0$ with $j \to \infty$, such that for every $\hbar = \hbar_j$:

1. There exists a space $\mathcal{H}_\hbar$ of *finite dimension*, an operator $\hat{\mathcal{F}}_\hbar : \mathcal{H}_\hbar \to \mathcal{H}_\hbar$ which is *quasi unitary* in the sense that there exists $\varepsilon_\hbar \geq 0$ with $\varepsilon_{\hbar_j} \to 0$, with $j \to \infty$ and

$$\forall u \in \mathscr{H}_{\hbar}, (1 - \varepsilon_{\hbar}) \|u\| \leq \left\| \hat{\mathscr{F}}_{\hbar} u \right\| \leq (1 + \varepsilon_{\hbar}) \|u\|. \tag{2.121}$$

2. The operator $\hat{\mathscr{F}}_{\hbar}$ satisfies the *asymptotic Gutzwiller Trace formula* for large time; i.e. there exists $0 < \theta < 1$ independent on $\hbar$ and some $C_{\hbar} > 0$ which may depend on $\hbar$, such that for $\hbar$ small enough (such that $\theta < 1 - \varepsilon_{\hbar}$):

$$\forall n \in \mathbb{N}, \quad \left| \mathrm{Tr}\left(\hat{\mathscr{F}}_{\hbar}^{n} \right) - \sum_{x = f^{n}(x)} \frac{e^{iS_{x,n}/\hbar}}{\sqrt{\left| \det\left(1 - Df_{x}^{n} \right) \right|}} \right| \leq C_{\hbar} \theta^{n}. \tag{2.122}$$

Let us notice first that Theorem 2.12 (for the case $V = V_0$) provides a solution to Problem 2.1: this is the quantum operator $\hat{\mathscr{F}}_{\hbar} : \mathscr{H}_{\hbar} \to \mathscr{H}_{\hbar}$ defined in (2.73) obtained with the choice of potential $\tilde{V} = \tilde{V}_0$, giving $V = V_0$. Indeed (2.121) holds true and (2.122) holds true from (2.120) and because $\theta := r_1^+ + \varepsilon < 1$.

Some importance of the Gutzwiller trace formula (2.122) comes from the following property which shows uniqueness of the solution to the problem:

Proposition 2.9. *If $\hat{\mathscr{F}}_{\hbar} : \mathscr{H}_{\hbar} \to \mathscr{H}_{\hbar}$ is a solution of Problem 2.1 then the spectrum of $\hat{\mathscr{F}}_{\hbar}$ is uniquely defined (with multiplicities). In particular $\dim(\mathscr{H}_{\hbar})$ is uniquely defined.*

Proof. This is consequence of the following lemma.

Lemma 2.6. *If A, B are matrices and for any $n \in \mathbb{N}$, $|Tr(A^n) - Tr(B^n)| < C\theta^n$ with some $C > 0$, $\theta \geq 0$ then A and B have the same spectrum with same multiplicities on the spectral domain $|z| > \theta$.*

Proof. From the formula[20]:

$$\det(1 - \mu A) = \exp\left(-\sum_{n \geq 1} \frac{\mu^n}{n} \mathrm{Tr}(A^n) \right).$$

---

[20]This formula is easily proved by using eigenvalues λ_j of A and the Taylor series of $\log(1 - x) = -\sum_{n \geq 1} \frac{x^n}{n}$ which converges for $|x| < 1$:

$$\det(1 - \mu A) = \prod_j (1 - \mu \lambda_j) = \exp\left(\sum_j \log(1 - \mu \lambda_j) \right)$$

$$= \exp\left(-\sum_j \sum_{n \geq 1} \frac{(\mu \lambda_j)^n}{n} \right) = \exp\left(-\sum_{n \geq 1} \frac{\mu^n}{n} \mathrm{Tr}(A^n) \right).$$

The sum on the right is convergent if $1/|\mu| > \|A\|$. Notice that we have (with multiplicities): μ is a zero of $d_A(\mu) = \det(1 - \mu A)$ if and only if $z = \frac{1}{\mu}$ is a (generalized) eigenvalue of A. Using the formula we get that if $1/|\mu| > \theta$ then

$$
\left| \frac{\det(1 - \mu A)}{\det(1 - \mu B)} \right| \leq \exp \left(\sum_{n \geq 1} \frac{|\mu|^n}{n} \left| \mathrm{Tr}(A^n) - \mathrm{Tr}(B^n) \right| \right)
$$

$$
< \exp \left(C \sum_{n \geq 1} \frac{(|\mu| \theta)^n}{n} \right) = (1 - \theta |\mu|)^{-C} =: \mathcal{B}.
$$

Similarly $\left| \frac{\det(1 - \mu A)}{\det(1 - \mu B)} \right| > \frac{1}{\mathcal{B}}$, hence $d_A(\mu)$ and $d_B(\mu)$ have the same zeroes on $1/|\mu| > \theta$. Equivalently A and B have the same spectrum on $|z| > \theta$.   □

If $\hat{G}_\hbar$ is another solution of Problem 2.1 then (2.122) implies $\left| \mathrm{Tr}\left(\hat{\mathscr{F}}_\hbar^n \right) - \mathrm{Tr}\left(\hat{G}_\hbar^n \right) \right| \leq 2C\theta^n$ and Lemma 2.6 tells us that $\hat{G}_\hbar$ and $\hat{\mathscr{F}}_\hbar$ have the same spectrum on $|z| > \theta$. But by hypothesis (2.121) their spectrum is in $|z| > 1 - \varepsilon_\hbar > \theta$. Therefore all their spectrum coincides. This finishes the proof of Proposition 2.9.   □

Remark 2.53. Previous results in the literature concerning the "semiclassical Gutzwiller formula" for "quantum maps" do not provide an answer to the Problem 2.1 above. We explain why. For any reasonable quantization of the Anosov map $f : M \to M$, e.g. the Weyl quantization or geometric quantization, one obtains a family of unitary operators $\hat{\mathscr{F}}_\hbar : \mathscr{H}_\hbar \to \mathscr{H}_\hbar$ acting in some finite dimensional (family of) Hilbert spaces. So this answer to (2.121). Using semiclassical analysis it is possible to show a Gutzwiller formula like (2.122) but with an error term on the right hand side of the form $\mathcal{O}(\hbar \theta^n)$ with $\theta = e^{h_{top}/2} > 1$ where $h_{top} > 0$ is the topological entropy which represents the exponential growing number of periodic orbits ([21] and references therein). Using more refined semiclassical analysis at higher orders, the error can be made

$$
\mathcal{O}\left(\hbar^M \theta^n \right) \tag{2.123}
$$

with any $M > 0$ [21], but nevertheless one has a total error which gets large after the so-called *Ehrenfest time*: $n \gg M \frac{\log(1/\hbar)}{\lambda_0}$. So all these results obtained from any quantization scheme do not provide an answer to the Problem 2.1. We may regard the operator in (2.73) as the only "quantization procedure" for which (2.122) holds true. For that reason we may call it a *natural quantization* of the Anosov map f.

2.4.2 Gutzwiller Trace Formula for Contact Anosov Flows

These results are in a work in preparation [25]. These results are transposition of the results of Sect. 2.4.1 in the case of contact Anosov flow.

Fig. 2.27 Graph of the flow

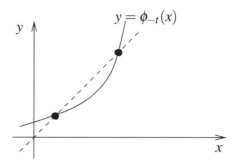

We write the transfer operator as

$$\left(\hat{F}_t v\right)(x) = \left(e^{tA}v\right)(x) = e^{\left(\int^t V\right)(x)} \cdot \left(v\left(\phi_{-t}(x)\right)\right)$$

$$= \int_M K_t(x, y) \, v(y) \, dy$$

with the distributional Schwartz kernel given by $K_t(x, y) = e^{\left(\int^t V\right)(x)} \delta\left(y - \phi_{-t}(x)\right)$ (this is the "graph of the flow"). For $t > 0$, the *"flat trace"* is:

$$\mathrm{Tr}^{\flat}\left(\hat{F}_t\right) := \int_M K_t(x, x) \, dx = \int_M e^{\int^t V} \cdot \delta\left(x - \phi_{-t}(x)\right) dx.$$

See Fig. 2.27.

As in Proposition 2.5 we obtain[21] the *"Atiyah-Bott trace formula"* as a sum over periodic orbits of the flow ϕ_t:

$$\mathrm{Tr}^{\flat}\left(\hat{F}_t\right) = \sum_{\gamma:o.p.} |\gamma| \sum_{n \geq 1} \frac{e^{\int^t V} \cdot \delta\left(t - n|\gamma|\right)}{\left|\det\left(1 - D_{(u,s)}\phi_{-t}(\gamma)\right)\right|} \tag{2.124}$$

with $|\gamma| > 0$: period of γ and n: number of repetitions. This is a distribution in $\mathscr{D}'(\mathbb{R}_t)$.

Question 2.2. Relation between the periodic orbits γ and the Ruelle spectrum of $A = -X + V$, generator of $\hat{F}_t = e^{tA}$?

---

[21] For this we use that if $f : \mathbb{R}^n \to \mathbb{R}^n$ with fixed point $f(0) = 0$, with the change of variable $y = f(x)$, we write $\int \delta(f(x)) \, dx = \frac{1}{|\det Df(0)|} \int \delta(y) \, dy = \frac{1}{|\det Df(0)|}$.

2.4.2.1   Zeta Function

- Observation: in linear algebra, the eigenvalues of a matrix $\mathbf{A}$ are zeroes of the holomorphic function[22]

$$\mathbf{d}(z) := \det(z - \mathbf{A})$$

$$= \mathbf{d}(z_0) \cdot \exp\left(\lim_{\varepsilon \to 0}\left[-\int_\varepsilon^\infty \frac{1}{t}e^{-zt}\mathrm{Tr}\left(e^{t\mathbf{A}}\right)dt\right]_{z_0}^{z}\right), \quad z_0 \notin \mathrm{Spec}(\mathbf{A}).$$

For $\mathrm{Re}(z) \gg 1$ we define the *"spectral determinant"* or *zeta function*:

$$d(z) := \exp\left(-\int_{|\gamma|_{min}}^\infty \frac{1}{t}e^{-zt}\mathrm{Tr}^\flat\left(e^{t\mathbf{A}}\right)dt\right)$$

$$\underset{(2.124)}{=} \exp\left(-\sum_\gamma \sum_{n \geq 1} \frac{e^{\int^t V} \cdot e^{-zn|\gamma|}}{n\left|\det\left(1 - D_{(u,s)}\phi_{n|\gamma|}(\gamma)\right)\right|}\right).$$

Theorem 2.13 ([29]). *For an Anosov vector field X, $d(z)$ has an analytic extension on $\mathbb{C}$. Its zeroes are Ruelle resonances with multiplicities.*

Remark 2.54. in 2008, Baladi-Tsujii [8] have a similar result for Anosov diffeomorphisms.

2.4.2.2   Application: Counting Periodic Orbits

The objective is to express in term of Ruelle spectrum the *counting function*:

$$\pi(T) := \sharp\{\gamma : periodic - orbit, \quad |\gamma| \leq T\} = \sum_{\gamma, |\gamma| \leq T} 1.$$

Observe that $\left|\det\left(1 - D_{(u,s)}\phi_t(\gamma)\right)\right|^{-1} \underset{t\infty}{\simeq} \det\left(D\phi_{t/E_u}\right)^{-1}$. The choice of potential $V = \mathrm{div}X_{/E_u}$ gives $e^{\int^t V} = \det\left(D\phi_{t/E_u}\right)$ and $e^{\int^t V}\left|\det\left(1 - D_{(u,s)}\phi_t(\gamma)\right)\right|^{-1} \simeq 1$.

Theorem 2.14 ([29] (with pinching hypothesis)). *There exists $\delta > 0$ s.t.*

$$\pi(T) = \mathrm{Ei}\left(h_{top}T\right) + O\left(e^{(h_{top}-\delta)T}\right) \underset{T \to \infty}{\sim} \frac{e^{h_{top}T}}{h_{top}T}$$

---

[22]Write $(z-A)^{-1} = \int_0^\infty e^{-(z-A)t}dt$, and $d(z) = \det(z-A) = \exp(\mathrm{Tr}(\log(z-A)))$ hence $\frac{d}{dz}\log d(z) = \mathrm{Tr}(z-A)^{-1} = \int_0^\infty e^{-zt}\mathrm{Tr}\left(e^{tA}\right)dt$.

with $\mathrm{Ei}\,(x) := \int_{x_0}^{x} \frac{e^y}{y}\,dy$ and h_{top} dominant eigenvalue of $A = -X + \mathrm{div}\,X_{/E_u(x)}$ called topological entropy.

2.4.2.3 Semiclassical Zeta Function

Observe that we have

$$\left|\det\left(1 - D_{(u,s)}\phi_t\,(\gamma)\right)\right|^{-1} \underset{t\infty}{\simeq} \det\left(D\phi_{t/E_u}\right)^{-1/2}\left|\det\left(1 - D_{(u,s)}\phi_t\,(\gamma)\right)\right|^{-1/2}$$

and $\det\left(D\phi_{t/E_u}\right)^{-1/2} = e^{-\frac{1}{2}\int^t \mathrm{div}\,X_{/E_u}}\,e^{-\frac{1}{2}\int^t V_0}$ so in (2.124) we have

$$e^{\int^t V}\left|\det\left(1 - D_{(u,s)}\phi_t\,(\gamma)\right)\right|^{-1} \underset{t\infty}{\simeq} e^{\int^t D}\left|\det\left(1 - D_{(u,s)}\phi_t\,(\gamma)\right)\right|^{-1/2}.$$

We define the *"Gutzwiller-Voros zeta function"* or *"semi-classical zeta function"* by

$$d_{G-V}\,(z) := \exp\left(-\sum_{\gamma}\sum_{n\geq 1} \frac{e^{-zn|\gamma|}e^{\int^t D}}{n\left|\det\left(1 - D_{(u,s)}\phi_{n|\gamma|}\,(\gamma)\right)\right|^{1/2}}\right). \qquad (2.125)$$

Theorem 2.15 ([25]). *The semiclassical zeta function* $d_{G-V}\,(z)$ *has an meromorphic extension on* $\mathbb{C}$. *On* $\mathrm{Re}\,(z) > \gamma_1^+$, $d_{G-V}\,(z)$ *has finite number of poles and its zeroes coincide (up to finite number) with the Ruelle eigenvalues of* A.

See Fig. 2.26. The motivation for studying $d_{G-V}\,(z)$ comes from the Gutzwiller semiclassical trace formula in quantum chaos. Also in the case of *surface with constant curvature*, and $V = V_0 = \frac{1}{2}$, we have $D_{(u,s)}\phi_{n|\gamma|}\,(\gamma) = \begin{pmatrix} e^{|\gamma|n} & 0 \\ 0 & e^{-|\gamma|n} \end{pmatrix}$. This gives

$$d_{G-V}\,(z) \underset{(2.125)}{=} \exp\left(-\sum_{\gamma}\sum_{n\geq 1}\sum_{m\geq 0} \frac{1}{n}e^{-n|\gamma|(z+\frac{1}{2}+m)}\right)$$

$$= \prod_{\gamma}\prod_{m\geq 0}\left(1 - e^{-(z+\frac{1}{2}+m)|\gamma|}\right) =: \zeta_{Selberg}\left(z + \frac{1}{2}\right).$$

Proof. Put $x = e^{-|\gamma|n}$ and use that

$$\left|\det\left(1 - \begin{pmatrix} 1 - x^{-1} & 0 \\ 0 & 1 - x \end{pmatrix}\right)\right|^{-1/2} = x^{1/2}\,(1-x)^{-1} = x^{1/2}\sum_{m\geq 0} x^m.$$

Fig. 2.28 Zeroes of $\zeta_{Selberg}$

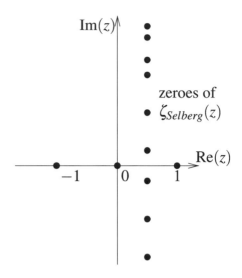

Therefore $d_{G-V}(z)$ *"generalizes" the Selberg zeta function* $\zeta_{Selberg}$ for case of variable curvature (or contact Anosov flows). Compare Fig. 2.28 with Fig. 2.26.

2.5 Appendix: Some Definitions and Theorems of Semiclassical Analysis

2.5.1 Class of Symbols

Notations: For $x \in \mathbb{R}^n$, $\langle x \rangle := \sqrt{1 + |x|^2}$ and we use the standard multi-indices notation $\partial_x^\alpha f := \frac{\partial^{\alpha_i} f}{\partial x_i^{\alpha_i}} \ldots \frac{\partial^{\alpha_n} f}{\partial x_n^{\alpha_n}}$.

2.5.1.1 Symbols with Constant Order

The following classes of symbols have been introduced by Hörmander [33]. Let M be a smooth compact manifold.

Definition 2.10. Let $\mu \in \mathbb{R}$ called the *order*. Let $0 \leq \delta < \frac{1}{2} < \rho \leq 1$. The *class of symbols* $S_{\rho,\delta}^\mu$ contains smooth functions $p \in C^\infty (T^*M)$ such that on any charts of $U \subset M$ with coordinates $x = (x_1, \ldots x_n)$ and associated dual coordinates $\xi = (\xi_1, \ldots \xi_n)$ on T_x^*U, any multi-index $\alpha, \beta \in \mathbb{N}^n$, there is a constant $C_{\alpha,\beta}$ such that

$$\left| \partial_\xi^\alpha \partial_x^\beta p(x, \xi) \right| \leq C_{\alpha,\beta} \, \langle \xi \rangle^{\mu - \rho|\alpha| + \delta|\beta|} . \tag{2.126}$$

The case $\rho = 1, \delta = 0$ is very common. We denote $S^\mu := S^\mu_{1,0}$.

For example on a chart, $p(x, \xi) = \langle \xi \rangle^\mu$ is a symbol $p \in S^\mu$.

If $\mu \leq \mu'$ then $S^\mu \subset S^{\mu'}$. We have $S^{-\infty} := \bigcap_{\mu \in \mathbb{R}} S^\mu = \mathscr{S}(T^*M)$.

2.5.1.2   Symbols with Variable Order in T^*M

We refer to [26, Section A.2.2] for a precise description of theorems related to symbols with variable orders. This class of symbols is useful for Anosov diffeomorphisms and Anosov flows on a manifold. Let M be a smooth compact manifold.

Definition 2.11. Let $m(x, \xi) \in S^0_{1,0}$ be a real-valued called *variable order* and let $0 \leq \delta < \frac{1}{2} < \rho \leq 1$. The *class of symbols* $S^{m(x,\xi)}_{\rho,\delta}$ contains smooth functions $p \in C^\infty(T^*M)$ such that on any charts of $U \subset M$ with coordinates $x = (x_1, \ldots x_n)$ and associated dual coordinates $\xi = (\xi_1, \ldots \xi_n)$ on $T^*_x U$, any multi-index $\alpha, \beta \in \mathbb{N}^n$, there is a constant $C_{\alpha,\beta}$ such that

$$|\partial^\alpha_\xi \partial^\beta_x p(x, \xi)| \leq C_{\alpha,\beta} \langle \xi \rangle^{m(x,\xi) - \rho|\alpha| + \delta|\beta|}. \qquad (2.127)$$

Example 2.6. For example $A(x, \xi) = \langle \xi \rangle^{m(x,\xi)}$ in (2.51) belongs to $S^{m(x,\xi)}_{\rho,\delta}$ with any $0 < \delta < \frac{1}{2} < \rho < 1$.

2.5.1.3   Symbols with Variable Order in $\mathbb{R}^{2d}$

Here we introduce a class of symbol specifically for application to Sect. 2.3.1 on $\mathbb{R}^{2d}$. We denote $z = (x, \xi) \in \mathbb{R}^{2d}$.

Definition 2.12. Let $\mu \in \mathbb{R}$ and $0 < \rho \leq 1$. A symbol $p(z) \in S^\mu_\rho$ is a function $p \in C^\infty(\mathbb{R}^{2d})$ such that $\forall \alpha \in \mathbb{N}^{2d}, \exists C_\alpha > 0$

$$|\partial^\alpha_z p(z)| \leq C_\alpha \langle z \rangle^{\mu - \rho|\alpha|}. \qquad (2.128)$$

Example 2.7. Example: $m(z)$ after Eq. (2.34) belongs to $S^0 := S^0_1$.

Definition 2.13. Let $m(z) \in S^0$. The *class of symbols* $S^{m(z)}_\rho$ with variable order $m(z)$ contains smooth functions $p \in C^\infty(\mathbb{R}^{2d})$ such that $\forall \alpha \in \mathbb{N}^{2d}, \exists C_\alpha > 0$,

$$|\partial^\alpha_z p(z)| \leq C_\alpha \langle z \rangle^{m(z) - \rho|\alpha|}. \qquad (2.129)$$

Example 2.8. A_C (z) in Eq. (2.34) belongs to $S_\rho^{m(z)}$ with any $0 < \rho < 1$.

Proof. Let us observe: we have $\partial_x A = (\partial_x m) \log \langle \xi \rangle . A$ but $(\partial_x m) \in S^0$ and $\log \langle \xi \rangle \in S^\varepsilon$ for every $\varepsilon > 0$ so $\partial_x A \in S^\varepsilon$. We have

$$\partial_\xi A = \left((\partial_\xi m) \log \langle \xi \rangle + m . \frac{\partial_\xi \langle \xi \rangle}{\langle \xi \rangle} \right) . A$$

but $(\partial_\xi m) \in S^{-1}$, $\log \langle \xi \rangle \in S^\varepsilon$ for any $\varepsilon > 0$, $m \in S^0$, $\partial_\xi \langle \xi \rangle \in S^0$, $\langle \xi \rangle^{-1} \in S^{-1}$ so $\partial_\xi A \in S^{m-\rho}$ with $\rho = 1 - \varepsilon$. $\square$

2.5.2 Pseudo-differential Operators (PDO)

2.5.2.1 Quantization

"Quantization" is a map Op which maps a symbol p to an operator Op (p) with specific properties. For example, its inverse maps the algebra of operators (for the composition) to an algebra on the symbols which coincide with the ordinary product of functions at first order.

Definition 2.14. If $p \in S_{\rho,\delta}^m (T^*M)$ is a symbol with order m, its *standard quantization* is the operator Op (p) : $\mathscr{D}'(M) \to \mathscr{D}'(M)$, $C^\infty(M) \to C^\infty(M)$ whose distribution kernel is smooth outside the diagonal and such that on a local coordinate chart $U \subset \mathbb{R}^n$, it is given up to a smoothing operator by

$$(\text{Op}\,(p)\,u)\,(x) := \frac{1}{(2\pi)^n} \iint e^{i(x-y)\cdot\xi}\, p(x,\xi)u(y)dyd\xi. \tag{2.130}$$

We say that Op (p) is a *pseudo-differential operator (PDO)* with ordinary symbol p:

- For example if X is a vector field on M, the operator $\hat{p} = \text{Op}\,(p) = -iX$ is a PDO with ordinary symbol

$$p(x,\xi) = X(\xi). \tag{2.131}$$

- For example on $M = \mathbb{R}^d$, if $p(x,\xi) = \sum_{\alpha \in \mathbb{N}^d} p_\alpha(x)\xi^\alpha$ (with a finite number of terms) then Op (p) is the differential operator:

$$\text{Op}\,(p)\,u = \sum_{\alpha \in \mathbb{N}^d} p_\alpha(x)(-i\partial_x)^\alpha u.$$

Definition 2.15. For *Weyl quantization*, (2.130) is replaced by [42, (14.5), p. 60]:

$$(\mathrm{Op}_W\,(p)\,u)\,(x) := \frac{1}{(2\pi)^n}\iint e^{i(x-y)\cdot\xi}\,p\left(\frac{x+y}{2},\xi\right)u\,(y)\,dyd\xi. \qquad (2.132)$$

We say that $\mathrm{Op}_W\,(p)$ is a *pseudo-differential operator* or *PDO* with Weyl symbol p.

Remark 2.55. Weyl quantization is often preferred other standard quantization because it has specific interesting properties. First a real symbol $p \in S^m\,(M),m \in \mathbb{R}$, is quantized in a formally self-adjoint operator $\hat{P} = \mathrm{Op}\,(p)$. Secondly, a change of coordinate systems preserving the volume form changes the symbol at a subleading order $S^{\mu-2}$ only. In other words, on a manifold with a fixed smooth density dx, the Weyl symbol p of a given pseudodifferential operator $\hat{P}$ is well defined modulo terms in $S^{\mu-2}$.

For example if X is a vector field on M, the operator $\hat{p} = -iX$ is a PDO with Weyl symbol

$$p_W\,(x,\xi) = X\,(\xi) + \frac{i}{2}\mathrm{div}\,(X) \qquad (2.133)$$

Indeed from [42, (14.7), p. 60], in a given chart where $X = \sum X_j\,(x)\frac{\partial}{\partial x^j} \equiv X\,(x)\,\partial_x$,

$$p_W\,(x,\xi) = \exp\left(\frac{i}{2}\partial_x\partial_\xi\right)(X\,(x)\,.\xi) = X\,(x).\xi + \frac{i}{2}\partial_x X = X\,(\xi) + \frac{i}{2}\mathrm{div}\,(X)$$

and $\mathrm{div}\,(X)$ depends only on the choice of the volume form, see [41, p. 125]. Notice that this symbol does not depend on the choice of coordinates systems provided the volume form is expressed by $dx = dx_1 \ldots dx_n$. The first term $p_0\,(x,\xi) = X\,(\xi)$ in (2.133) belongs to S^1 is called the *principal symbol* of $\hat{p}$. The second term $\frac{i}{2}\mathrm{div}\,(X)$ in (2.133) belongs to S^0 and is called the *subprincipal symbol* of $\hat{p}$.

2.5.2.2 Composition

Theorem 2.16 ([42, Prop. (3.3), p. 11] "Composition of PDO"). *If* $A \in S^{m_1}_{\rho,\delta}$ *and* $B \in S^{m_2}_{\rho,\delta}$ *then* $\mathrm{Op}\,(A)\,\mathrm{Op}\,(B) = \mathrm{Op}\,(AB) + \mathcal{O}\left(\mathrm{Op}(S^{m_1+m_2-(\rho-\delta)}_{\rho,\delta})\right)$, *i.e. the symbol of* $\mathrm{Op}\,(A)\,\mathrm{Op}\,(B)$ *is the product* AB *and belongs to* $S^{m_1+m_2}_{\rho,\delta}$ *modulo terms in* $S^{m_1+m_2-(\rho-\delta)}_{\rho,\delta}$.

We also have:

Theorem 2.17 ([42, Eq. (3.24)(3.25), p. 13]). *The symbol of the commutator* $[\mathrm{Op}\,(A),\mathrm{Op}\,(B)]$ *is the Poisson bracket* $-i\,\{A,B\}$ *modulo* $S^{m_1+m_2-2(\rho-\delta)}_{\rho,\delta}$. *The symbol* $-i\,\{A,B\}$ *belongs to* $S^{m_1+m_2-(\rho-\delta)}_{\rho,\delta}$.

We also recall [41, (10.8), p. 43] that $\{A, B\} = -\mathbf{X}_B(A)$ *where* $\mathbf{X}_B$ *is the Hamiltonian vector field generated by* B.

2.5.2.3 Bounded and Compact PDO

For PDO with order zero we have:

Theorem 2.18 (**"L^2 continuity theorem"**). *Let* $p \in S_\rho^0$. *Then* $\mathrm{Op}(p)$ *is a bounded operator and for any* $\varepsilon > 0$ *there is a decomposition* $\mathrm{Op}(p) = \hat{p}_\varepsilon + \hat{K}_\varepsilon$ *with* $\hat{K}_\varepsilon \in \mathrm{Op}(S^{-\infty})$ *smoothing operator,* $\|\hat{p}_\varepsilon\| \leq L + \varepsilon$ *and*

$$L = \limsup_{(x,\xi) \in T^*M} |p(x,\xi)|.$$

For PDO with negative order we have:

Theorem 2.19. *Let* $p \in S_\rho^\mu$ *with* $\mu < 0$ *then* $\mathrm{Op}(p)$ *is a compact operator. If* $\mu < -d$ *so that* $\int_{T^*M} |p(x,\xi)| \, dx d\xi < \infty$, *then* $\mathrm{Op}(p)$ *is a trace class operator and*

$$\mathrm{Tr}(\mathrm{Op}(p)) = \frac{1}{(2\pi)^d} \int p(x,\xi) \, dx d\xi.$$

2.5.3 Wavefront

The wavefront set of a distribution has been introduced by Hörmander. The wavefront set corresponds to the directions in T^*X where the distribution is not C^∞ (i.e. the local Fourier transform is not rapidly decreasing). The wavefront set of a PDO is the directions in T^*X where the symbol is not rapidly decreasing:

Definition 2.16 ([30, p. 77], [42, p. 27]). *If* $(x_0, \xi_0) \in T^*M \setminus 0$, *we say that* $A \in S^m$ *is non characteristic (or elliptic) at* (x_0, ξ_0) *if* $\left| A(x,\xi)^{-1} \right| \leq C |\xi|^{-m}$ *for* (x,ξ) *in a small conic neighborhood of* (x_0, ξ_0) *and* $|\xi|$ *large. If* $u \in \mathscr{D}'(M)$ *is a distribution, we say that* u *is* C^∞ *at* $(x_0, \xi_0) \in T^*X \setminus 0$ *if there exists* $A \in S^m$ *non characteristic (or elliptic) at* (x_0, ξ_0) *such that* $(\mathrm{Op}(A) u) \in C^\infty(M)$. *The wavefront set of the distribution* u *is*

$$\mathrm{WF}(u) := \{(x_0, \xi_0) \in T^*M \setminus 0, \quad u \text{ is not } C^\infty \text{ at } (x_0, \xi_0)\}$$

The *wavefront set of the operator* $\mathrm{Op}(A)$ is the smallest closed cone $\Gamma \subset T^*M \setminus 0$ such that $A_{/\complement\Gamma} \in S^{-\infty}(\complement\Gamma)$.

2.6   Some General References (Books or Reviews)

- On dynamical systems: [5, 10, 34].
- On semiclassical analysis: [30, 36, 42, 47].
- On quantum chaos: [31, 37].

Acknowledgements We thank the scientific committee and the organizers of the INdAM Workshop "Geometric, Analytic and Probabilistic approaches to dynamics in negative curvature", (F.Ledrappier, C.Liverani, G.Mondello, F.Dal'Bo, M.Peigné, A.Sambusetti) where these lectures were given by the first author.

References

1. V.I. Arnold, A. Avez, *Méthodes ergodiques de la mécanique classique* (Gauthier Villars, Paris, 1967)
2. V.I. Arnold, *Les méthodes mathématiques de la mécanique classique* (Ed. Mir, Moscou, 1976)
3. M.F. Atiyah, R. Bott, A Lefschetz fixed point formula for elliptic differential operators. Bull. Am. Math. Soc. **72**, 245–250 (1966)
4. M.F. Atiyah, R. Bott, A Lefschetz fixed point formula for elliptic complexes. I. Ann. Math. (2) **86**, 374–407 (1967)
5. V. Baladi, *Positive Transfer Operators and Decay of Correlations* (World Scientific, Singapore, 2000)
6. V. Baladi, M. Tsujii, Dynamical determinants and spectrum for hyperbolic diffeomorphisms, in *Probabilistic and Geometric Structures in Dynamics*, ed. by K. Burns, D. Dolgopyat, Ya. Pesin. Contemporary Mathematics, Volume in honour of M. Brin's 60th birthday (American Mathematical Society, Providence, RI, 2008). arxiv:0606434
7. V. Baladi, M. Tsujii, Anisotropic Hölder and Sobolev spaces for hyperbolic diffeomorphisms. Ann. Inst. Fourier **57**, 127–154 (2007)
8. V. Baladi, M. Tsujii, Dynamical determinants and spectrum for hyperbolic diffeomorphisms, in *Geometric and Probabilistic Structures in Dynamics*. Volume 469 of Contemporary Mathematics (American Mathematical Society, Providence, 2008), pp. 29–68
9. M. Blank, G. Keller, C. Liverani, Ruelle-Perron-Frobenius spectrum for Anosov maps. Nonlinearity **15**, 1905–1973 (2002)
10. M. Brin, G. Stuck, *Introduction to Dynamical Systems* (Cambridge University Press, Cambridge, 2002)
11. O. Butterley, C. Liverani, Smooth Anosov flows: correlation spectra and stability. J. Mod. Dyn. **1**(2), 301–322 (2007)
12. A. Cannas Da Salva, *Lectures on Symplectic Geometry* (Springer, Berlin/Heidelberg, 2001)
13. K. Datchev, S. Dyatlov, M. Zworski, Sharp polynomial bounds on the number of ruelle resonances. Ergod. Theory Dynam. Syst. 1–16 (Cambridge University Press, Cambridge, 2012)
14. D. Dolgopyat, On decay of correlations in Anosov flows. Ann. Math. (2) **147**(2), 357–390 (1998)
15. D. Dolgopyat, On mixing properties of compact group extensions of hyperbolic systems. Israel J. Math. **130**, 157–205 (2002)
16. D. Dolgopyat, M. Pollicott, Addendum to 'periodic orbits and dynamical spectra'. Erg. Theory Dyn. Syst. **18**(2), 293–301 (1998)
17. M. Dimassi, J. Sjöstrand, *Spectral Asymptotics in the Semi-classical Limit*. Volume 268 of London Mathematical Society Lecture Note Series (Cambridge University Press, Cambridge, 1999)

18. S. Dyatlov, Resonance projectors and asymptotics for r-normally hyperbolic trapped sets (2013). arXiv preprint arXiv:1301.5633v2
19. S. Dyatlov, M. Zworski, Dynamical zeta functions for anosov flows via microlocal analysis (2013). arXiv preprint arXiv:1306.4203
20. F. Faure, Prequantum chaos: resonances of the prequantum cat map. J. Mod. Dyn. **1**(2), 255–285 (2007). arXiv:nlin/0606063
21. F. Faure, Semiclassical formula beyond the ehrenfest time in quantum chaos.(I) trace formula. Annales de l'Institut Fourier, No.7. **57**, 2525–2599 (2007)
22. F. Faure, J. Sjöstrand, Upper bound on the density of ruelle resonances for anosov flows. a semiclassical approach. Commun. Math. Phys. Issue 2 **308**, 325–364 (2011). arXiv:1003.0513v1
23. F. Faure, M. Tsujii, Prequantum transfer operator for symplectic anosov diffeomorphism. Asterisque (2012, submitted). Arxiv preprint arXiv:1206.0282
24. F. Faure, M. Tsujii, Band structure of the ruelle spectrum of contact anosov flows. C. R. Math. **351**, 385–391 (2013). arXiv preprint arXiv:1301.5525
25. F. Faure, M. Tsujii, The semiclassical zeta function for geodesic flows on negatively curved manifolds (2013). arXiv preprint arXiv:1311.4932
26. F. Faure, N. Roy, J. Sjöstrand, A semiclassical approach for anosov diffeomorphisms and ruelle resonances. Open Math. J. **1**, 35–81 (2008). arXiv:0802.1780
27. L. Flaminio, G. Forni, Invariant distributions and time averages for horocycle flows. Duke Math. J. **119**(3), 465–526 (2003)
28. P. Foulon, B. Hasselblatt, Contact anosov flows on hyperbolic 3-manifolds. Geom. Topol. **17**(2), 1225–1252 (2013). Preprints, Tufts University
29. P. Giulietti, C. Liverani, M. Pollicott, Anosov flows and dynamical zeta functions. Ann. Math. **178**(2), 687–773 (2012). arXiv:1203.0904
30. A. Grigis, J. Sjöstrand, *Microlocal Analysis for Differential Operators*. Volume 196 of London Mathematical Society Lecture Note Series (Cambridge University Press, Cambridge, 1994). An introduction.
31. M. Gutzwiller, *Chaos in Classical and Quantum Mechanics* (Springer, New York, 1991)
32. B. Helffer, J. Sjöstrand, *Résonances en limite semi-classique* (Resonances in semi-classical limit). Memoires de la S.M.F. 24/25 (1986)
33. L. Hörmander, *The Analysis of Linear Partial Differential Operators. III*. Volume 274 of Grundlehren der Mathematischen Wissenschaften (Fundamental Principles of Mathematical Sciences) (Springer, Berlin, 1985). Pseudodifferential operators
34. A. Katok, B. Hasselblatt, *Introduction to the Modern Theory of Dynamical Systems* (Cambridge University Press, Cambridge, 1995)
35. C. Liverani, On contact Anosov flows. Ann. Math. (2) **159**(3), 1275–1312 (2004)
36. A. Martinez, *An Introduction to Semiclassical and Microlocal Analysis*. Universitext (Springer, New York, 2002)
37. S. Nonnenmacher, Some open questions in 'wave chaos'. Nonlinearity **21**(8), T113–T121 (2008)
38. S. Nonnenmacher, M. Zworski, Decay of correlations for normally hyperbolic trapping (2013). arXiv:1302.4483
39. H.H. Rugh, The correlation spectrum for hyperbolic analytic maps. Nonlinearity **5**(6), 1237–1263 (1992)
40. J. Sjöstrand, Asymptotic distribution of eigenfrequencies for damped wave equations. Publ. Res. Inst. Math. Sci **36**(5), 573–611 (2000)
41. M. Taylor, *Partial Differential Equations*, vol. I (Springer, New York, 1996)
42. M. Taylor, *Partial Differential Equations*, vol. II (Springer, New York, 1996)
43. L.N. Trefethen, M. Embree, *Spectra and Pseudospectra* (Princeton University Press, Princeton, 2005)
44. M. Tsujii, Quasi-compactness of transfer operators for contact anosov flows. Nonlinearity **23**(7), 1495–1545 (2010). arXiv:0806.0732v2 [math.DS]

45. M. Tsujii, Contact anosov flows and the fourier–bros–iagolnitzer transform. Erg. Theory Dyn. Syst. **32**(06), 2083–2118 (2012)
46. S. Zelditch, Quantum maps and automorphisms, in *The Breadth of Symplectic and Poisson Geometry*. Volume 232 of Progress in Mathematics (Birkhäuser Boston, Boston, 2005), pp. 623–654
47. M. Zworski, *Semiclassical Analysis*, vol. 138 (American Mathematical Society, Providence, 2012)

Index

η-Hölder continuous, 42, 50
$\mathbb{Z}^d$-fibers, 47

Action, 74, 122
Anisotropic
    mixing property, 54
    Sobolev space, 88, 93, 107, 111
Anosov, 69, 75
    one form, 75
Atiyah-Bott trace formula, 94, 125

Band structure, 98
Birkhoff sum, 115
Burger, 28

Canonical
    Euler vector field, 76
    map, 103
Casimir operator, 113
Cat map, 69
Central limit theorem, 72
Class of symbols, 128, 129
Clt
    along subsequences, 26
    for vector valued functions, 26
Cohomologous, 9
Composition of PDO, 131
Connection one form, 73
Contact
    Anosov flow, 76, 77
    one form, 73
    vector field, 76
Convolution, 40, 42
Correlation function, 71, 79, 80

Cotangent space, 67
Counting function, 126
Curvature, 73

Damping function, 97, 115
Decay of correlation, 68
Diagonal flow, 45
Donsker invariance principle, 20
Doob inequality, 21

Egorov Theorem, 87
Ehrenfest time, 124
Equidistribution of unstable leaves, 19, 28, 31, 39
Ergodic, 72
Escape function, 86, 92, 110
Evolution of functions, 66
Exponentially mixing, 71, 79

Filtration, 10, 13
Flat trace, 94, 125
Fourier
    integral operators, 120
    mode, 71
Frame bundle, 33

Geodesic flow, 26, 32, 76
    with negative curvature, 76
Geometric
    prequantization, 72
    quantization, 101
Gordin's method, 9
Gutzwiller trace formula, 122
Gutzwiller-Voros zeta function, 127

F. Dal'Bo et al. (eds.), *Analytic and Probabilistic Approaches to Dynamics in Negative Curvature*, Springer INdAM Series 9, DOI 10.1007/978-3-319-04807-9,
© Springer International Publishing Switzerland 2014

Holonomy, 51
Horocyclic flows, 27
Hyperbolic automorphism on the torus,
        69

Jan's method, 54

L2 continuity Theorem, 132
Lagrangian, 72
Liouville one form, 76
Lyapounov function, 86

Markov partition, 14
Martingale, 6

Natural quantization, 122
Negative sectional curvature, 77
Non characteristic (or elliptic), 132
Non wandering set, 85

Order, 128
    function, 86

Perron-Frobenius operator, 81, 88, 90
Potential, 96
    function, 106
Prequantum
    line bundle, 97
    map, 73
Pseudo-differential operator (PDO), 93, 130

Quantum
    correlation function, 117
    operator, 100
    space, 100

Recurrence, 47
Reeb vector field, 76
Resonance, 108
Reversed martingale, 6

Ruelle
    operator, 81
    spectrum, 83
Ruelle-Pollicott resonances, 83, 91, 107

Scattering on the trapped set, 85
Selberg zeta function, 128
Semiclassical functional calculus, 111
Semi-classical zeta function, 127
Spectral determinant, 126
Stable subspace, 69
Standard quantization, 130
Statistical properties, 72
Structural stability, 70
Symplectic manifold, 72
Symplectization, 118

Theorem of composition of PDO, 87
Theorem of L^{2} -continuity, 87
Tight, 21
Topological
    entropy, 127
    pressure, 96, 114
Transfer operator, 81, 106
Trapped set, 68, 85, 110

Unstable
    Jacobian, 96, 114
    subspace, 69

Variable order, 92, 129

Wavefront set
    of the distribution, 132
    of the operator, 132
Wave packet, 67
Weyl
    formula, 98
    law, 116
    quantization, 131

Zeta function, 126